DIGITAL COMPUTER
METHODS IN
ENGINEERING

DIGITAL COMPUTER METHODS IN ENGINEERING

SHAHEN A. HOVANESSIAN

Hughes Aircraft Company, Culver City, California

LOUIS A. PIPES

University of California, Los Angeles

McGRAW-HILL BOOK COMPANY   New York   St. Louis

San Francisco   London   Sydney   Toronto   Mexico   Panama

To Mary, Linda, and Christina

PREFACE

This book contains introductory material on numerical methods of solving engineering problems by digital computers. The book is written for engineers and scientists who desire to achieve a working knowledge of digital computer calculations and programming methods for subsequent application in their field of practice. For this reason, the book covers basic digital computer methods which can be applied to all fields of engineering, rather than specific applications. The level of the contents is such as to make the book particularly suited for self-study. Each chapter can be taken independently for the study of a specific subject or a specific application.

The book can be used as a junior-senior level text for a one semester course in digital methods given to engineering students from various branches of engineering or as a reference book to supplement the numerical computation portion of general applied mathematics courses.

It is assumed that the readers have had two years of engineering mathematics. No knowledge of computer programming is assumed on the part of the reader. Digital computer programs are given for a number of numerical examples to familiarize the reader with actual computer programming logic. These programs are written in FORTRAN (used by general purpose computers) and BASIC (used by time-sharing computers) programming languages. These programming languages are so close to actual algebraic statements that, with the explanatory paragraphs given with each program, the reader will have no difficulty following the logic of the computer programs. It should be noted that the digital computer programs are written primarily for simplicity of logic rather than efficiency of computations.

The following is a general description of the contents of various chapters.

Chapters 1 and 2 contain discussions and methods of numerical solution of determinants, matrices, and linear algebraic equations. Included in these chapters are methods of evaluating characteristic values and vectors using digital computers. A physical interpretation of characteristic values and vectors is given at the end of the second chapter.

Chapter 3 includes a number of interpolation formulas and methods of curve fitting primarily used with statistical data. Polynomial root finding methods are also described in this chapter.

Chapter 4 gives a discussion of time-frequency domain analysis using Fourier series and transforms and describes the use of digital computers in spectral analysis of time dependent functions.

Chapters 5, 6, and 7 start by describing the basic relations between the differential and difference operators and methods of reducing a differential equation to a difference equation. Subsequent discussions deal with solution of various types of differential equations using difference methods and digital computers. An appendix containing elementary solutions of typical differential equations is included for readers desiring a review of classical methods of solution of differential equations.

Chapter 8 describes the numerical methods of solution of partial differential equations. Specifically, the numerical solutions of Laplace's equation, the heat equation, and the wave equation are presented.

Chapters 9 and 10 give a discussion of the elements of linear and dynamic programming. On the assumption that the majority of readers are not familiar with the subjects of these chapters, these chapters are written at a somewhat lower level than the rest of the text.

Most of the material described above is taken from lecture notes of applied mathematics courses offered at UCLA over the past several years. In addition, a number of computational methods currently used in industry are included. For this reason the authors wish to acknowledge the valuable suggestions and comments of their students and their colleagues during the preparation of this work. The editorial assistance of Mrs. Rosalie Kenniston of Hughes Aircraft Company in the preparation of the original manuscript is also gratefully acknowledged.

S. A. Hovanessian
Louis A. Pipes

CONTENTS

1 DETERMINANTS, MATRICES, AND LINEAR SIMULTANEOUS EQUATIONS

The theory of determinants is intimately connected with that of matrices and with the solution of simultaneous linear algebraic equations. A great deal of the terminology and some of the operations involved in matrix algebra are based on a knowledge of elementary determinant theory. In this chapter we start with a brief discussion of the elements of the theory of determinants. This elementary discussion should provide an adequate review for the reader who is primarily interested in matrix algebra and matrix calculus from a standpoint of applications to engineering and physics. The rest of the chapter is devoted to matrices and numerical solutions of matrix equations and linear simultaneous equations.

DETERMINANTS AND SIMULTANEOUS EQUATIONS

It is sometimes convenient to express the combination of terms $(ad-bc)$ in the following form:

$$D = \begin{vmatrix} a & c \\ b & d \end{vmatrix} = ad-bc \tag{1.1}$$

D is called a determinant of the second order. The letters a, b, c, d, are called the elements of the determinant; $(ad-bc)$ is called its expansion. The first row of D consists of the elements a and c. The second row of D consists of the elements b and d. The elements a and b comprise the first column and the elements c and d the second column.

A determinant of the third order has three rows and three columns; a determinant of the order n has n rows and n columns. The *transpose* of a determinant D of any order will be denoted by D'; it is the determinant formed by interchanging the rows and columns of D.

It is therefore evident that if

$$D = \begin{vmatrix} a & c \\ b & d \end{vmatrix}, \quad \text{then} \quad D' = \begin{vmatrix} a & b \\ c & d \end{vmatrix} \tag{1.2}$$

Determinants of the nth Order

An nth order determinant A is a square array of n^2 quantities a_{ij} called the elements of the determinant written in the form

$$A = \begin{vmatrix} a_{11} & a_{12} & a_{13} \cdots a_{1n} \\ a_{21} & a_{22} & a_{23} \cdots a_{2n} \\ \cdots \cdots \cdots \cdots \cdots \cdots \cdots \\ a_{n1} & a_{n2} & a_{n3} \cdots a_{nn} \end{vmatrix} \tag{1.3}$$

Consider the following definitions of quantities with respect to the above determinant.

Minors. If in the determinant A of Eq. (1.3) we delete the ith row and the jth column and form a determinant from all the remaining elements, we shall have a new determinant of $(n-1)$ rows and columns.

This new determinant is defined to be the *minor* of the element a_{ij}. The minor of the element a_{ij} will be denoted by M_{ij}.

Cofactors. The cofactor of an element of a determinant a_{ij} is the minor M_{ij} of that element with a sign attached to it. The sign is determined by the numbers i and j which fix the position of a_{ij} in the determinant A. The sign is chosen by the equation

$$C_{ij} = (-1)^{i+j} M_{ij} \tag{1.4}$$

where C_{ij} is the *cofactor* of the element a_{ij} and M_{ij} is the minor of the element a_{ij}.

In the general nth order case if C_{ij} is the cofactor of any element a_{ij} in the determinant A, the determinant can be expressed as the sum of the elements in any one row or any one column each multiplied by its cofactor. That is,

$$D = \sum_j a_{ij} C_{ij} \quad \text{(any row } i) \tag{1.5}$$

$$D = \sum_i a_{ij} C_{ij} \quad \text{(any column } j) \tag{1.6}$$

By repeating the expansion represented by Eq. (1.5) or (1.6), a determinant of large order n can be expressed in terms of determinants of steadily smaller order and hence can be evaluated.

It is easy to demonstrate that, in general,

$$\sum_{i=1}^{n} a_{ji} C_{jk} = \sum_{i=1}^{n} a_{ij} C_{kj} = \begin{cases} D & \text{if } i = k \\ 0 & \text{if } i \neq k \end{cases} \tag{1.7}$$

It should be noted that the evaluation of determinants in terms of cofactors as given by Eqs. (1.5) and (1.6) involves excessive numbers of arithmetical operations. It can be shown that by using this method, the number of multiplications can be approximated by $n!$ where n is the order of the determinant. For a determinant of order 10 this will result in 10! or 3.6×10^6 multiplications. In succeeding sections, methods using the fundamental properties of determinants will be given where the number of multiplications is reduced to $n^3/3$, or about 300 multiplications for evaluating a 10th order determinant.

Fundamental Properties of Determinants

From the basic formulas (1.5) and (1.6), the following properties of determinants can be deduced:

1. If all the elements of a row or of a column are zero, the determinant is equal to zero. This may be seen by expanding in terms of the elements of that row or column in which the elements are equal to zero.

2. If all elements but one in a row or column are zero, the determinant is equal to the product of that element and its cofactor.

3. The value of a determinant is not altered when the rows are changed to columns and the columns to rows. In other words, the determinant and its transpose are equal.

4. The interchange of any two columns or two rows of a determinant changes the sign of the determinant.

5. If two rows or two columns of a determinant are identical, the determinant is equal to zero.

6. If all the elements in any row or any column are multiplied by any factor, the determinant is multiplied by that factor.

7. If each element in any column or any row of a determinant is expressed as the sum of two quantities, the determinant can be expressed as the sum of two determinants of the same order.

8. It is possible, without changing the value of a determinant, to multiply the elements of any row (or any column) by the same constant and add the products to any other row (or column).

Use of the Fundamental Rules in Evaluating a Determinant

The fundamental rules enumerated above may be used in the numerical evaluation of determinants. For example, let it be required to evaluate the determinant

$$
D = \begin{vmatrix} 2 & -1 & 5 & 1 \\ 1 & 4 & 6 & 3 \\ 4 & 2 & 7 & 4 \\ 3 & 1 & 2 & 5 \end{vmatrix} \tag{1.8}
$$

This determinant will be greatly simplified if we use the above properties to make all its elements except one in some row or column equal to zero. The presence of the factor -1 in the second column suggests that we add four times the first row to the second, then add two times the first row to the third row, and finally add the first row to the fourth row. By principle 8 these operations do not change the value of the determinant. Hence we have

$$D = \begin{vmatrix} 2 & -1 & 5 & 1 \\ 9 & 0 & 26 & 7 \\ 8 & 0 & 17 & 6 \\ 5 & 0 & 7 & 6 \end{vmatrix} = \begin{vmatrix} 9 & 26 & 7 \\ 8 & 17 & 6 \\ 5 & 7 & 6 \end{vmatrix} \tag{1.9}$$

We now subtract the elements of the last column from the first column of the third-order determinant of Eq. (1.9) and thus obtain

$$D = \begin{vmatrix} 2 & 26 & 7 \\ 2 & 17 & 6 \\ -1 & 7 & 6 \end{vmatrix} \tag{1.10}$$

We now add two times the third row to the first row and two times the third row to the second row and thus obtain

$$D = \begin{vmatrix} 0 & 40 & 19 \\ 0 & 31 & 18 \\ -1 & 7 & 6 \end{vmatrix} = - \begin{vmatrix} 40 & 19 \\ 31 & 18 \end{vmatrix} \tag{1.11}$$

Now we may subtract the second row from the first row and obtain

$$D = - \begin{vmatrix} 9 & 1 \\ 31 & 18 \end{vmatrix} = -(162 - 31) = -131 \tag{1.12}$$

The procedure described above is much shorter than a direct application of Eq. (1.5) or (1.6).

Chió Method for Evaluating Determinants

The above procedure can be systematically applied to reduce the order of large determinants for numerical evaluation. In matrix algebra this procedure is referred to as the method of Chió. Consider the 4×4 determinant

$$D = \begin{vmatrix} a_{11} & a_{12} & a_{13} & a_{14} \\ a_{21} & a_{22} & a_{23} & a_{24} \\ a_{31} & a_{32} & a_{33} & a_{34} \\ a_{41} & a_{42} & a_{43} & a_{44} \end{vmatrix} \tag{1.13}$$

The idea of this method is to reduce the order of the determinant from 4 to 3 and from 3 to 2, and so on. This results in a 1×1 determinant which will give the actual value of the original determinant. To describe this, we proceed as follows. Make any element of the determinant, say element (1, 1), equal to unity by dividing the first row through by a_{11}. This results in

$$D = a_{11} \begin{vmatrix} 1 & \dfrac{a_{12}}{a_{11}} & \dfrac{a_{13}}{a_{11}} & \dfrac{a_{14}}{a_{11}} \\ a_{21} & a_{22} & a_{23} & a_{24} \\ a_{31} & a_{32} & a_{33} & a_{34} \\ a_{41} & a_{42} & a_{43} & a_{44} \end{vmatrix} \tag{1.14}$$

Denoting $a'_{12} = a_{12}/a_{11}$, $a'_{13} = a_{13}/a_{11}$, $a'_{14} = a_{14}/a_{11}$ and multiplying the first row by a_{21} and subtracting from the second row, then multiplying the first row by a_{31} and subtracting from the third row and, finally, multiplying the first row by a_{41} and subtracting from the fourth row, we have

$$D' = a_{11} \begin{vmatrix} 1 & a'_{12} & a'_{13} & a'_{14} \\ 0 & a_{22} - a_{21}a'_{12} & a_{22} - a_{21}a'_{13} & a_{24} - a_{21}a'_{14} \\ 0 & a_{32} - a_{31}a'_{12} & a_{33} - a_{31}a'_{13} & a_{34} - a_{31}a'_{14} \\ 0 & a_{42} - a_{41}a'_{12} & a_{43} - a_{41}a'_{13} & a_{44} - a_{41}a'_{14} \end{vmatrix} \tag{1.15}$$

The value of this determinant now becomes

$$D'' = a_{11}(-1)^{1+1} \begin{vmatrix} a''_{11} & a''_{12} & a''_{13} \\ a''_{21} & a''_{22} & a''_{23} \\ a''_{31} & a''_{32} & a''_{33} \end{vmatrix} \tag{1.16}$$

where $(-1)^{1+1}$ is the sign of the element $(1, 1)$ and double primes are used to denote elements

$$a''_{11} = a_{22} - a_{21}a'_{12}, \quad a''_{12} = a_{22} - a_{21}a'_{13}, \cdots$$
$$a''_{21} = a_{32} - a_{31}a'_{12}, \quad a''_{22} = a_{33} - a_{31}a'_{13}, \cdots \tag{1.17}$$

and so on. Thus Eq. (1.16) is a determinant of order 3×3. Repeating the above operation, this determinant will reduce to a 2×2 and subsequently to a 1×1 with multipliers $a_{11}a''_{11}a'''_{11}\cdots$. Thus, the value of the original 4×4 determinant can be obtained.

In cases where any of pivotal elements— $a_{11}, a'_{11}, a''_{11}, \cdots$ in the above derivation—become zero, the appropriate columns or rows should be interchanged to obtain a nonzero pivotal element. The process can then be continued. If an element which is already unity exists in a certain row, that element can be taken as the pivotal element, thereby reducing the number of divisions required to obtain a unity pivotal element. This, however, in a digital computer program will complicate the logic and will require additional logical statements.

The total number of arithmetical operations* required in evaluating determinants by the method of Chió can be shown to be

$$\frac{(n - 1)(2n^2 - n + 6)}{6} \quad \text{multiplications}$$

$$\frac{n(n - 1)}{2} \quad \text{divisions} \tag{1.18}$$

*Multiplications and divisions rather than additions and subtractions are used as a basis for calculating the required digital computer computation time, since multiplications and divisions take far more (by a factor of 5 to 10) computation time in digital computers.

where n is the order of the determinant to be evaluated. For a 10×10 determinant the above relations give, approximately, 275 multiplications and 45 divisions. Assuming that each operation will take 50 microseconds on a digital computer, the total computation time will be $320 \times 50 \times 10^{-3} = 16$ milliseconds.

FORTRAN Program for Determinant Evaluation

The logic of evaluating determinants by the method of Chió, as discussed above, can be programmed for digital computer solution. Such a program, written in FORTRAN language, is given below.

```
CSAH1               DETERMINANT EVALUATION PROGRAM
   10      DIMENSION S(100,100)
   20      DATA  N/3/
   30      DATA ((S(I,J),J=1,3),I=1,3)/1.,2.,3.,3.,4.,6.,2.,1.,1./
   40      L=1
   50      K=2
   60      XM1=1.
   70      XM=S(L,L)
   80      DO 95 J=L,N
   90      S(L,J)=S(L,J)/XM
   95      CONTINUE
  100      DO 140 I=K,N
  110      X=S(I,L)
  120      DO 140 J=L,N
  130      S(I,J)=S(I,J)−S(L,J)*X
  140      CONTINUE
  150      L=L+1
  160      K=K+1
  170      XM1=XM1*XM
  180      IF (L−N)70,190,190
  190      XM1=XM1*S(N,N)
  200      WRITE (2,210) XM1
  210      FORMAT (22H VALUE OF DETERMINANT= 1PE10.3)
  220      STOP
  230      END
```

The computer solution obtained is

VALUE OF DETERMINANT = 1.000E00

In order to completely understand the above program the reader should consult the FORTRAN reference manual listed as [1]. The logic of the program, however, can be explained by considering the evaluation of the following 3×3 determinant used in the program. It is required to evaluate the determinant

$$\begin{vmatrix} 1 & 2 & 3 \\ 3 & 4 & 6 \\ 2 & 1 & 1 \end{vmatrix} \tag{1.19}$$

Statement 10 of the program will reserve computer memory space for a 100×100 determinant. Statements 20 and 30 will read into the computer memory the order of the determinant $N = 3$ and the elements of the determinant $S(I, J)$, starting from the first row and proceeding to the second and the third rows. Statements 40 through 60 will set indices and the value of $XM1 = 1.0$ to be used later in the program. Statement 70 will set XM equal to the value of element $(1, 1)$ or $S(1, 1)$ of the determinant or in this case $S(1, 1) = XM = 1.0$. The loop consisting of statements 80 through 95 will divide the elements of the first row by XM. In this case, since XM is equal to 1, determinant of Eq. (1.19) will be unchanged. The double loop consisting of statements 100 through 140 and 120 through 140 will multiply the first row by element $S(2, 1)$ and subtract the resulting first row from the second row and will repeat this process with the third row. At the end of this operation, the resulting determinant will be

$$\begin{vmatrix} 1 & 2 & 3 \\ 0 & -2 & -3 \\ 0 & -3 & -5 \end{vmatrix} \tag{1.20}$$

Statements 150 and 160 set in indices $L = L + 1$ and $K = K + 1$, resulting in $L = 2$ and $K = 3$. The value of $XM1$ is equal to previous $XM1$ times XM, or at this point $XM1 = 1.0$. Thus, at statement 170 the determinant is reduced to a second-order determinant

$$(1) X \begin{vmatrix} -2 & -3 \\ -3 & -5 \end{vmatrix} \tag{1.21}$$

Statement 180 checks to see if the value of L is smaller than the order of the determinant, in which case the above operation starting with statement 70 is repeated. The second application of logic described by statements 70 through 170 will result in

$$(1)(-2) \begin{vmatrix} 1 & 1.5 \\ & \\ 0 & -.5 \end{vmatrix} \tag{1.22}$$

This time, since N will be equal to L at statement 180, the logic will go to statement 190. This will set

$$XM1 = (1)(-2)(-.5) = 1 \tag{1.23}$$

which is the value of the determinant as printed in the computer solution. Note that in the above program the 3×3 determinant was kept in computer storage while its elements were changed and operated upon without manipulating the location of unused elements.

The Solution of Linear Equations; Cramer's Rule

A general system of n simultaneous linear equations can be written in the form

$$\begin{aligned} a_{11}x_1 + a_{12}x_2 + \cdots + a_{1n}x_n &= k_1 \\ a_{21}x_1 + a_{22}x_2 + \cdots + a_{2n}x_n &= k_2 \\ \cdots\cdots\cdots\cdots\cdots\cdots\cdots\cdots\cdots\cdots\cdots\cdots\cdots \\ a_{n1}x_1 + a_{n2}x_2 + \cdots + a_{nn}x_n &= k_n \end{aligned} \tag{1.24}$$

The solution of the system (1.24) can be shown to be given by

$$x_r = \frac{D_r}{D}, \quad r = 1, 2, 3, \ldots, n \tag{1.25}$$

where D is the determinant of the coefficients of Eqs. (1.24)

$$
D = \begin{vmatrix}
a_{11} & a_{12} & a_{13} & \cdots & a_{1n} \\
a_{21} & a_{22} & a_{23} & \cdots & a_{2n} \\
\cdots & \cdots & \cdots & \cdots & \cdots \\
a_{n1} & a_{n2} & a_{n3} & \cdots & a_{nn}
\end{vmatrix}
\tag{1.26}
$$

The determinant D_r is obtained by replacing the rth column of D by $(k_1, k_2, k_3, \ldots, k_n)$. The result (1.25) is called "Cramer's Rule" in the mathematical literature. It presents an elegant but laborious method for solving sets of simultaneous equations.

From Eq. (1.18) for large values of n, high-order determinants, the number of multiplications can be approximated by $n^3/3$ and divisions by $n^2/2$. Assuming the number of divisions to be negligible as compared to the number of multiplications, the number of arithmetical operations in evaluating nth order determinants by the method of Chió can be set at $n^3/3$. Using Cramer's rule in solving n simultaneous equations in n unknowns, n + 1 determinants should be evaluated. Using Chió's method for the evaluation of these determinants will result in the total number of arithmetical operations $(n^3/3) (n + 1) = n^4/3$. In succeeding sections it will be shown that Gauss' method of elimination requires only $n^3/3$ multiplications in solving n simultaneous equations in n unknowns. For this reason Cramer's rule should be taken more for its historical and academic value rather than for its value in practical application.

The Regular Case. If the determinant of the coefficients D of the equations does not vanish so that $D \neq 0$, the set of equations is said to be regular. In this case they have the unique solution (1.25).

The Singular Case. If $D = 0$, the set of equations is said to be singular. A singular set of equations may or may not be consistent. The question of consistency will not be considered here. However, it may be pointed out that in simple cases the question of consistency can be decided by trial.

Homogeneous Equations in n Unknowns

If all the numbers $k_1, k_2, k_3, \ldots, k_n$ in Eqs. (1.24) are zero, the set of equations is said to be a set of homogeneous equations. It can be proven that the necessary and sufficient condition that a set of homogeneous equations should have nonzero solutions (x_1, x_2, x_3, ..., x_n not all zero) is that the determinant of the coefficients D should vanish, or $D = 0$.

An important relation between the ratios of the unknowns in the case of nontrivial solutions can be established in the following heuristic manner. Let it be assumed, for the moment, that the determinant D is not zero and that only one of the k's in Eqs. (1.24), say k_i, is different from zero. In this case by Cramer's rule we would have

$$x_r = \frac{C_{ir}k_i}{D} \tag{1.27}$$

In this special case, the ratio of any two unknowns is given by

$$\frac{x_r}{x_s} = \frac{C_{ir}}{C_{is}} \quad \text{for} \quad i = 1, 2, 3, \ldots, n \tag{1.28}$$

The result (1.28) is independent of the values of both k_i and the determinant D. It may be inferred, therefore, that Eq. (1.28) holds also when both k_i and D are zero. The correctness of this conclusion can be demonstrated in a rigorous fashion. It is therefore interesting to note that when the set of homogeneous equations has nontrivial solutions, the ratios of the unknowns are fixed by Eq. (1.28) in terms of the ratios of cofactors of the determinant D.

MATRIX ALGEBRA, LINEAR SIMULTANEOUS EQUATIONS

Matrices provide a compact and flexible notation that is particularly useful in studying linear transformations, and present an organized method for the solution of systems of linear algebraic and linear differential equations. Since many engineering problems are formulated in terms of linear algebraic or linear differential equations, it is natural that their solution will be expedited by the use of matrix algebra and matrix calculus. Matrix algebra has been found so useful in pure and applied mathematics that it has been termed "the arithmetic of higher mathematics."

Definition and Notation

A table of n by m quantities, called elements, arranged in a rectangular array of n rows and m columns is called a matrix with n rows and m columns. If a_{ij} is the element of the ith row and the jth column, then the matrix A can be written in the following manner:

$$A = \begin{bmatrix} a_{11} & a_{12} & a_{13} \cdots a_{1m} \\ a_{21} & a_{22} & a_{23} \cdots a_{2m} \\ \cdots\cdots\cdots\cdots\cdots\cdots\cdots \\ a_{n1} & a_{n2} & a_{n3} \cdots a_{nm} \end{bmatrix} \tag{1.29}$$

The matrix A of Eq. (1.29) is said to be a matrix of order (n, m). If the number n of rows is equal to the number m of columns, then the matrix A is called a *square* matrix.

Besides square matrices, two other general types of matrices frequently occur. One is the *row* matrix R:

$$R = [r_1 \quad r_2 \quad r_3 \cdots r_n] \tag{1.30}$$

The other is the *column* matrix C

$$C = \begin{bmatrix} c_1 \\ c_2 \\ c_3 \\ c_n \end{bmatrix} \tag{1.31}$$

Elementary Operations on Matrices

Transposition. It is possible to interchange the rows and the columns of the matrix A of Eq. (1.29) to obtain a new matrix A' of the form

$$A' = \begin{bmatrix} a_{11} & a_{21} & a_{31} \cdots a_{n1} \\ a_{12} & a_{22} & a_{32} \cdots a_{n2} \\ \cdots\cdots\cdots\cdots\cdots\cdots\cdots \\ a_{1m} & a_{2m} & a_{3m} \cdots a_{nm} \end{bmatrix} \tag{1.32}$$

The matrix A' is called the *transpose* of A. The transpose of the row matrix Eq. (1.30), R', is a column, and the transpose of the column Eq. (1.31), C', is a row matrix.

Addition of Matrices. Let A and B be two matrices of the same order, that is, they have the same number of rows and columns. Let

$$A = [a_{ij}], \quad B = [b_{ij}] \tag{1.33}$$

Then by the sum $A + B$ of the matrices A and B we shall mean the uniquely obtainable matrix C whose elements c_{ij} are obtained by the equation

$$c_{ij} = a_{ij} + b_{ij}, \quad \begin{matrix} i = 1, 2, 3, \ldots, n \\ j = 1, 2, 3, \ldots, m \end{matrix}, \quad C = A + B \tag{1.34}$$

In other words, to add two matrices of the same order, we calculate the matrix whose elements are precisely the numerical sum of the corresponding elements of the two given matrices. The addition of two matrices of different orders is not defined.

Equality of Matrices. To complete the preliminary definitions, we must make clear what is meant when two matrices are said to be equal. Two matrices A and B are defined to be equal provided that they have the same order and that their corresponding elements are equal. That is,

$$A = B \quad \text{if} \quad a_{ij} = b_{ij} \tag{1.35}$$

From the above definitions of the sum of two matrices and the equality of two matrices it follows that addition of matrices is associative, that is,

$$A + (B + C) = (A + B) + C = A + B + C \tag{1.36}$$

Before proceeding with the definition of multiplication of matrices, a word or two must be said about two very important special square matrices. One is the *zero* matrix—a matrix all of whose elements are zero, for example,

$$0 = \begin{bmatrix} 0 & 0 & 0 & 0 \\ 0 & 0 & 0 & 0 \\ 0 & 0 & 0 & 0 \end{bmatrix}, \quad \text{a zero matrix of order } (3, 4) \tag{1.37}$$

Another matrix of great importance is the *unit* matrix. This is a square matrix of any order whose principal diagonal elements are unity and has zeros for all its other elements; it is denoted usually by the letter I.
For example, the unit matrix of the fourth order is

$$I = \begin{bmatrix} 1 & 0 & 0 & 0 \\ 0 & 1 & 0 & 0 \\ 0 & 0 & 1 & 0 \\ 0 & 0 & 0 & 1 \end{bmatrix} \tag{1.38}$$

Multiplication of Matrices

Multiplication of a Matrix by a Scalar. By definition, the multiplication of a matrix A by an ordinary number or scalar k is effected by multiplying each element of the matrix A by the number k, thus obtaining a new matrix whose elements are ka_{ij}. That is, by definition,

$$B = kA \quad \text{where} \quad b_{ij} = ka_{ij} \tag{1.39}$$

For example,

$$3 \begin{bmatrix} 2 & 4 \\ 3 & 1 \end{bmatrix} = \begin{bmatrix} 6 & 12 \\ 9 & 3 \end{bmatrix} \tag{1.40}$$

The Multiplication of a Matrix by Another Matrix. The definition of multiplication of one matrix by another has been chosen in order to facilitate operations involving linear transformations by the use of matrix algebra.

The rule of multiplication is such that two matrices A and B can be multiplied together in the order AB only when the number of columns of A is equal to the number of rows of B. That is, if A is a matrix of order (m, n) and B is a matrix of order (p, q), the product AB is not defined unless $n = p$. If $n = p$, the matrices A and B are said to be *conformable*. In matrix algebra only conformable matrices can be multiplied together.

The product of a matrix A of order (m, p) by a matrix B of order (p, n) is defined to be the matrix C given by

$$C = AB = [c_{ij}] \tag{1.41}$$

where the elements c_{ij} of the matrix C are obtained by the equation

$$c_{ij} = \sum_{k=1}^{k=p} a_{ik} b_{kj} \tag{1.42}$$

It can be seen that the product matrix C is a matrix of order (m, n). The fact that a matrix of order (m, p), when multiplied by a matrix of order (p, n), results in a matrix of order (m, n) may be expressed symbolically in the following manner:

$$(m, p)(p, n) = (m, n) \tag{1.43}$$

This symbolic equation is useful in determining the order of the resulting matrix when several matrices are multiplied together. As an example of the definition of the rule of multiplication Eq. (1.42), let

$$A = \begin{bmatrix} a_{11} & a_{12} & a_{13} \\ a_{21} & a_{22} & a_{23} \\ a_{31} & a_{32} & a_{33} \end{bmatrix} \quad B = \begin{bmatrix} b_{11} & b_{12} \\ b_{21} & b_{22} \\ b_{31} & b_{32} \end{bmatrix} \tag{1.44}$$

In this case we have matrices of orders $(3, 3)$ and $(3, 2)$. They are therefore conformable and the product matrix C is of order $(3, 2)$. In this case the typical element c_{ij} of the product $C = AB$ is given by Eq. (1.42) in the form

$$c_{ij} = a_{i1} b_{1j} + a_{i2} b_{2j} + a_{i3} b_{3j} \tag{1.45}$$

The elements Eq. (1.45) are now arranged in the rectangular array,

$$
C = \begin{bmatrix}
(a_{11}b_{11} + a_{12}b_{21} + a_{13}b_{31}) & (a_{11}b_{12} + a_{12}b_{22} + a_{13}b_{32}) \\
(a_{21}b_{11} + a_{22}b_{21} + a_{23}b_{31}) & (a_{21}b_{12} + a_{22}b_{22} + a_{23}b_{32}) \\
(a_{31}b_{11} + a_{32}b_{21} + a_{33}b_{31}) & (a_{31}b_{12} + a_{32}b_{22} + a_{33}b_{32})
\end{bmatrix}
$$

$$(1.46)$$

where C is the product matrix AB. It can be seen that the result Eq. (1.46) can be obtained by taking the sum of continued products of the rows of A times the corresponding columns of B.

Properties of Matrix Multiplication. In general, matrix multiplication is not commutative, that is,

$$AB \neq BA \qquad (1.47)$$

This can be seen directly from the definition Eq. (1.42). The matrices, Eq. (1.44), are not conformable in the order BA and cannot be multiplied in this order.

In matrix algebra it is necessary to distinguish between *premultiplication,* as when A is premultiplied by B to give the product BA, and *postmultiplication,* as when A is postmultiplied by B to give the product AB. If we have the equality

$$AB = BA \qquad (1.48)$$

then the matrices A and B are said to *commute* or to be *permutable.* The unit matrix I, it can be noted, commutes with any square matrix of the same order. This is,

$$IA = AI \qquad (1.49)$$

provided that I and A are of the same order.

Continued Products of Matrices. Except for the fact that matrices do not commute in multiplication, all the ordinary laws of scalar algebra apply to the multiplication of matrices. Of particular importance is the associative law of continued products:

$$ABC = A(BC) = (AB)C \qquad (1.50)$$

The fact that the associative law is satisfied permits one to dispense with parentheses and to write ABC without ambiguity. It must be noted, however, that the product of a chain of matrices will have meaning only if the adjacent matrices of the chain are conformable.

The Reversal Law of Transposed Products. It can be shown as a consequence of the definition of the rule of matrix multiplication Eq. (1.42) that the transpose of a product of matrices is equal to the product of their transposes in reverse order. That is, if we have

$$M = AB \quad M' = B'A' \quad \text{(reversed order)} \tag{1.51}$$

where primes indicate transpose matrices. Similarly, if

$$M = ABC \quad \text{then} \quad M' = C'B'A' \tag{1.52}$$

This result is called the reversal law of transposed products.

Partitioned Matrices and Partitioned Multiplication

It is sometimes convenient for purposes of computation to extend the use of the fundamental laws of combinations of matrices to the case in which a matrix is considered to be constructed from elements that are themselves matrices. For example, consider the matrix A

$$A = \begin{bmatrix} a_{11} & a_{12} & | & a_{13} \\ a_{21} & a_{22} & | & a_{23} \\ - - - - - & + - - \\ a_{31} & a_{32} & | & a_{33} \end{bmatrix} = \begin{bmatrix} P & Q \\ R & S \end{bmatrix} \tag{1.53}$$

The matrix A above has been partitioned by the introduction of a vertical and a horizontal line into the four submatrices $P, Q, R,$ and S given by

$$P = \begin{bmatrix} a_{11} & a_{12} \\ a_{21} & a_{22} \end{bmatrix} \quad Q = \begin{bmatrix} a_{13} \\ a_{23} \end{bmatrix} \quad \begin{matrix} R = [a_{31}\, a_{32}] \\ S = a_{33} \end{matrix}$$

In this case, the matrix A has been partitioned in such a manner that the diagonal submatrices P and S are square, and the partitioning is diagonally symmetrical. Now let B be another square matrix of the third order partitioned similarly to A in the form

$$B = \begin{bmatrix} b_{11} & b_{12} & | & b_{13} \\ b_{21} & b_{22} & | & b_{23} \\ \overline{b_{31}} & \overline{b_{32}} & | & \overline{b_{33}} \end{bmatrix} = \begin{bmatrix} P_1 & Q_1 \\ R_1 & S_1 \end{bmatrix} \tag{1.54}$$

Now it can easily be seen that the sum $C = A + B$ can be expressed in terms of submatrices in the form

$$C = A + B = \begin{bmatrix} (P + P_1) & (Q + Q_1) \\ (R + R_1) & (S + S_1) \end{bmatrix} \tag{1.55}$$

This indicates that when two matrices of the same order are partitioned similarly, their sum can be obtained by adding their various submatrices as if they were elements.

Multiplication in Terms of Submatrices. As a consequence of the fundamental rule for the multiplication of matrices, a rectangular matrix B can be premultiplied by another rectangular matrix A, provided that the two matrices are conformable, that is, when the number of rows of B equals the number of columns of A. Now if A and B are both partitioned into submatrices such that the grouping of columns in A agrees with the grouping of rows in B, it can be shown that the product AB can be obtained by treating the submatrices as ordinary elements and proceeding according to the general rule of multiplication.

In the case of the matrices A and B discussed above, the partitioning is such that the product $D = AB$ can be carried out by treating the submatrices of Eqs. (1.53) and (1.54) as if they were ordinary elements, thus obtaining

$$D = AB = \begin{bmatrix} (PP_1 + QR_1) & (PQ_1 + QS_1) \\ (RP_1 + SR_1) & (RQ_1 + SS_1) \end{bmatrix} \tag{1.56}$$

Matrix Division; the Inverse Matrix

Having defined the addition and the multiplication of matrices, we now consider an operation that is analogous to the operation of division in ordinary scalar algebra. In ordinary algebra, if the product of two quantities b and a equals unity, so that

$$ba = 1 \tag{1.57}$$

we say that b is the reciprocal of a and we write

$$b = \frac{1}{a} = a^{-1} \tag{1.58}$$

In matrix algebra the unit matrix I of Eq. (1.38) has properties similar to that of the number one in ordinary algebra since $AI = IA = A$ if A is a square matrix of the same order as I. If we now have two square matrices A and B of the same order and if they satisfy the equation

$$BA = I \tag{1.59}$$

we see that Eq. (1.59) is the matrix analog of Eq. (1.57); we then say that B is the inverse matrix of A and we write

$$B = A^{-1} \quad \text{and} \quad A^{-1}A = I \tag{1.60}$$

Equation (1.60) is the analog of the scalar Eq. (1.58).

It can be shown that the necessary and sufficient condition for the inverse of matrix A is that the matrix A be nonsingular, that is, $|A| \neq 0$. In order to find the inverse of matrix A, let us define the *adjoint* of A as the *transpose* of the matrix formed by replacing the elements of A by their cofactors C_{ij}. For a (3×3) matrix, this is written as

$$\text{adjoint } (A) = \text{adj } A = \begin{bmatrix} C_{11} & C_{21} & C_{31} \\ C_{12} & C_{22} & C_{32} \\ C_{13} & C_{23} & C_{33} \end{bmatrix} \tag{1.61}$$

On the strength of Eq. (1.7) we can write

$$(A)(\text{adj } A) = D[I] \tag{1.62}$$

where $D(= |A|)$ is the determinant of matrix A and $[I]$ is the unit matrix. From the above equations the inverse matrix can be written

$$\frac{\text{adj } (A)}{D} = \frac{[I]}{(A)} = (A)^{-1} \tag{1.63}$$

or

$$A^{-1} = \frac{\text{adj } A}{|A|} \tag{1.64}$$

In practice the inverse of matrix A is formed by obtaining the transpose of matrix (A') and subsequently obtaining the matrix of the cofactors of A'. Division of the matrix of the cofactors by the value of the determinant of $A(= |A|)$ will give the inverse matrix A^{-1}.

Although the above method of matrix inversion represents the classical approach, computationally it involves an excessive number of arithmetical operations. It has been shown that evaluation of large-order determinants by the method of Chió involves, approximately, $n^3/3$ multiplications, where n is the order of the determinant. Using this as a basis, evaluation of the inverse of an n by n matrix by the method of cofactors, described above, will involve evaluation of n^2 cofactors of order $(n - 1)$ by $(n - 1)$ and, in addition, the determinant of the matrix, resulting in $(n^2)(n - 1)^3/3 + n^3/3$ multiplications. For large n the above can be approximated by $n^5/3$ multiplications. In the following sections it will be shown that matrix inversion, for example, by the augmented matrix method, can be accomplished by as few as n^3 multiplications, a reduction by a factor $n^2/3$ from the number of multiplications required by the cofactor method of matrix inversion using the method of Chió in evaluation of the cofactors.

Gauss' Method of Elimination in Solving Simultaneous Equations

This method is probably the simplest approach to the solution of a series of simultaneous equations. It consists of eliminating one unknown at a time from the set of equations and proceeding with the remaining equations. This, at the end, will result in only one unknown and one equation. Working backward from this equation, one can solve for one unknown at a time until the solution is complete. Consider the set of equations

$$a_{11}x_1 + a_{12}x_2 + a_{13}x_3 + \cdots + a_{1n}x_n = b_1$$

$$a_{21}x_1 + a_{22}x_2 + a_{23}x_3 + \cdots + a_{2n}x_n = b_2$$

$$\dots\dots\dots\dots\dots\dots\dots\dots\dots\dots\dots\dots\dots\dots\dots\dots \quad (1.65)$$

$$a_{i1}n_1 + a_{i2}x_2 + a_{i3}x_3 + \cdots + a_{in}x_n = b_i$$

$$\dots\dots\dots\dots\dots\dots\dots\dots\dots\dots\dots\dots\dots\dots\dots\dots$$

$$a_{n1}n_1 + a_{n2}x_2 + a_{n3}x_3 + \cdots + a_{nn}x_n = b_n$$

Solving the first equation for x_1, we get

$$x_1 = \frac{b_1}{a_{11}} - \sum_{i=2}^{n} a_{1i}x_i$$

Substituting this value of x_1 in the second through nth equation, we obtain a set of $n-1$ equations in $n-1$ unknowns x_2 through x_n. This set can be written

$$a'_{22}x_2 + a'_{23}x_3 + \cdots a'_{2n}x_n = b'_2$$

$$\dots\dots\dots\dots\dots\dots\dots\dots\dots\dots\dots\dots\dots$$

$$a'_{i2}x_2 + a'_{i3}x_3 + \cdots a'_{in}x_n = b'_i \quad (1.66)$$

$$\dots\dots\dots\dots\dots\dots\dots\dots\dots\dots\dots\dots\dots$$

$$a'_{n2}x_2 + a'_{n3}x_3 + \cdots a'_{nn}x_n = b'_n$$

where primes denote the new coefficients. Now, solving the first equation for x_2 and substituting in the second through nth equations, we obtain a set of $n-2$ equations in $n-2$ unknowns x_3 through x_n. Continuing this operation will result in two equations in two unknowns

$$c'_{n-1,n-1}x_{n-1} + c'_{n-1,n}x_n = d'_{n-1}$$

$$c'_{n,n-1}x_{n-1} + c'_{n,n}x_n = d'_n \quad (1.67)$$

and finally in one equation and one unknown

$$c_{n,n}'' x_n = d_{n1}''$$ (1.68)

where c's and d's represent constants. Thus, Eq. (1.68) will give the value of x_n. Having x_n, Eq. (1.67) will give the value of x_{n-1}. Continuing this process will result in values of $x_{n-2}, x_{n-3}, \ldots, x_2$ and x_1.

It can be shown that the number of multiplications and divisions required in the above method is given by

$$\frac{n(n+1)(2n+1)}{6}$$ multiplications

$$n$$ divisions

where n is the number of simultaneous equations to be solved. For large n, the above values (neglecting divisions) can be approximated by $n^3/3$, a factor of n advantage over the solution by Cramer's method using Chió's method for the evaluation of the determinants.

Triangulation Method in the Solution of Simultaneous Equations

The triangulation method is often used in obtaining the solution of a set of simultaneous linear equations. This method primarily consists of systematically reducing the number of unknowns from each equation, and ending up with a single equation and a single unknown. One can solve this equation for the unknown and work backward to solve for all the rest of the unknowns one at a time. To describe this in equation form, consider the set of linear equations

$$a_{11}x_1 + a_{12}x_2 + \cdots + a_{1n}x_n = b_1$$
$$a_{21}x_1 + a_{22}x_2 + \cdots + a_{2n}x_n = b_2$$
$$\vdots$$
$$a_{i1}x_1 + a_{i2}x_2 + \cdots + a_{in}x_n = b_i$$ (1.69)
$$\vdots$$
$$a_{n1}x_1 + a_{n2}x_2 + \cdots + a_{nn}x_n = b_n$$

Divide through the first equation by a_{11}. Multiply this equation by a_{21} and subtract from the second equation. Multiply the first equation by a_{31} and subtract from the third equation. Follow this procedure up to the nth equation. This will result:

$$x_1 + a'_{12}x_2 + \cdots + a'_{1n}x_n = b'_1$$
$$a'_{22}x_2 + \cdots + a'_{2n}x_n = b'_2$$
$$\vdots$$
$$a'_{i2}x_2 + \cdots + a'_{in}x_n = b'_i \qquad (1.70)$$
$$\vdots$$
$$a'_{n2}x_2 + \cdots + a'_{nn}x_n = b'_n$$

where primes denote new coefficients.

Now divide the second equation by a'_{22} and subsequently multiply the resulting equation by $a'_{32}, a'_{42}, \ldots$ and subtract from the third, fourth, ..., equations, respectively. This operation will eliminate unknowns x_2's from all of the equations except the first and the second equations. Continuing this procedure will triangulate the set of equations resulting in

$$x_1 + c_{12}x_2 + c_{13}x_3 + \cdots + c_{1n}x_n = d_1$$
$$x_2 + c_{23}x_3 + \cdots + c_{2n}x_n = d_2$$

$$(1.71)$$

$$x_{n-1} + c_{n-1,n}x_n = d_{n-1}$$
$$x_n = d_n$$

where c's and d's are constants resulting from arithmetical operations. The last equation gives the value of x_n. This x_n can be substituted in the equation before last resulting in the value of x_{n-1}. Continuing this process, all of the unknowns can be obtained.

The triangulation method can be recognized as a systematic procedure to effect Gauss' elimination method of solving simultaneous equations.

Since the logical operations of the triangulation method are straightforward it can easily be programmed for digital computer solution. The number of multiplications and divisions required is approximately the same as that of Gauss' elimination method.

Example. Consider the solution of the set of equations in unknowns x, y, and z by the above method.

$$\begin{aligned} x + 2y + z &= 4 \\ 2x + y - z &= 5 \\ x - y + 2z &= -3 \end{aligned}$$

$$(1.72)$$

Multiplying the first equation by -2 and adding to the second and multiplying the first equation by -1 and adding to the third will result in

$$\begin{aligned} x + 2y + z &= 4 \\ -3y - 3z &= -3 \\ -3y + z &= -7 \end{aligned}$$

Dividing the second equation by -3 we get

$$\begin{aligned} x + 2y + z &= 4 \\ y + z &= 1 \\ -3y + z &= -7 \end{aligned}$$

Multiplying the second equation by 3 and adding to the third will result in

$$\begin{aligned} x + 2y + z &= 4 \\ y + z &= 1 \\ 4z &= -4 \end{aligned}$$

The last equation gives $z = -1$. Having this, the second equation gives $y = 2$ and finally the first gives $x = 1$.

Solution of Simultaneous Linear Equations
by Matrix Inversion

Consider the set of n simultaneous equations in n unknowns:

$$a_{11}x_1 + a_{12}x_2 + a_{13}x_3 + \cdots + a_{1n}x_n = b_1$$
$$a_{21}x_1 + a_{22}x_2 + a_{23}x_3 + \cdots + a_{2n}x_n = b_2$$
$$\vdots \qquad\qquad\qquad\qquad\qquad\qquad (1.73)$$
$$a_{i1}x_1 + a_{i2}x_2 + a_{i3}x_3 + \cdots + a_{in}x_n = b_i$$
$$\vdots$$
$$a_{n1}x_1 + a_{n2}x_2 + a_{n3}x_3 + \cdots + a_{nn}x_n = b_n$$

where a's are the constant coefficient, x's are the unknowns, and b's are constants. The above equation can be written in matrix form as follows:

$$[A](x) = (B) \qquad\qquad\qquad\qquad (1.74)$$

where $[A]$ is a square $n \times n$ matrix, and (x) and (B) are column matrices. Matrices $[A]$, (B), and (x) can be written

$$[A] = \begin{bmatrix} a_{11} & a_{12} & \cdots & a_{1n} \\ a_{21} & a_{22} & \cdots & a_{2n} \\ \vdots & & & \\ a_{n1} & a_{n2} & \cdots & a_{nn} \end{bmatrix} \quad (B) = \begin{bmatrix} b_1 \\ b_2 \\ \vdots \\ b_n \end{bmatrix} \quad (x) = \begin{bmatrix} x_1 \\ x_2 \\ \vdots \\ x_n \end{bmatrix} \qquad (1.75)$$

Premultiplying both sides of Eq. (1.74) by $[A]^{-1}$ results in

$$(x) = [A]^{-1}(B) \qquad\qquad\qquad\qquad (1.76)$$

Thus the solution can be obtained by evaluating the inverse matrix $[A]^{-1}$ and multiplying it by column matrix B.

In previous pages, numerical methods for the solution of simultaneous equations were described, where the number of multiplications required for solving a set of n equations in n unknowns was, approximately, $n^3/3$.

In the following pages matrix inversion methods, to be used in conjunction with the above method, will be described where n^3 operations are required to invert an $n \times n$ matrix. Thus, the method of Eq. (1.76) will involve three times more multiplications in inverting the matrix A than previous methods require in solving for a set of n unknowns. For this reason the method of Eq. (1.76) should be applied only when a set of x's for a set of B's are to be calculated. Under these conditions the inverse matrix A^{-1} can be reapplied to calculate new x's for new B's.

Augmented Matrix Method of Matrix Inversion

The augmented matrix method of matrix inversion is based, primarily, on the Gauss elimination method of solving simultaneous equations. For this reason its derivation will be based upon a set of n equations and n unknowns. It will be shown that the procedure of inverting matrices by this method can be mechanized without relating it to a set of simultaneous equations.

Consider the set of equations

$$a_{11}x_1 + a_{12}x_2 + \cdots + a_{1n}x_n = b_1$$

$$a_{21}x_1 + a_{22}x_2 + \cdots + a_{2n}x_n = b_2$$

$$\vdots$$

$$a_{n1}x_1 + a_{n2}x_2 + \cdots + a_{nn}x_n = b_n$$

$$(1.77)$$

It is desired to find the inverse of the matrix of the coefficients $[A]^{-1}$. This set of equations can be written in augmented matrix form as follows:

$$
\begin{bmatrix}
\overbrace{\qquad A \qquad} & \overbrace{\qquad A_1 \qquad} \\
a_{11} & a_{12} & a_{13} \cdots a_{1n} & 1 & 0 & 0 & \cdots & 0 \\
a_{21} & a_{22} & a_{23} \cdots a_{2n} & 0 & 1 & 0 & \cdots & 0 \\
\vdots \\
a_{i1} & a_{i2} & a_{i3} \cdots a_{in} & 0 & 0 & 0 \cdots 1 \cdots 0 \\
\vdots \\
a_{n1} & a_{n2} & a_{n3} \cdots a_{nn} & 0 & 0 & 0 & \cdots & 1
\end{bmatrix}
\begin{bmatrix}
x_1 \\
x_2 \\
\vdots \\
x_n \\
-b_1 \\
-b_2 \\
\vdots \\
-b_n
\end{bmatrix}
= 0
$$

$$(1.78)$$

where the large matrix is the augmented matrix of A. The idea of this method is to make all of the elements along the diagonal of the part A of the augmented matrix $(a_{11}, a_{22}, \ldots, a_{nn})$ equal to unity and the rest of the elements of part A equal to zero. In doing this the A_1 part of the matrix will fill up, resulting in

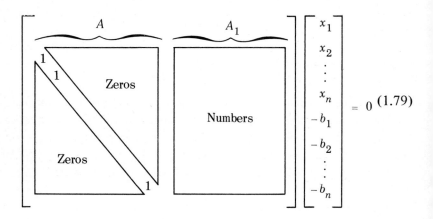

$$= 0 \quad (1.79)$$

Multiplying these matrices and transforming the (x) and (B) matrices to proper sides of the equal sign, we have

$$\begin{bmatrix} x_1 \\ \vdots \\ x_i \\ \vdots \\ x_n \end{bmatrix} = \begin{bmatrix} \overbrace{}^{A_1} \\ \text{Numbers} \end{bmatrix} \begin{bmatrix} b_1 \\ b_2 \\ \vdots \\ b_n \end{bmatrix} \qquad (1.80)$$

Comparing this to Eq. (1.76) we see that $[A]^{-1} = A_1$. Substituting the values of b's from the given set of linear equations and multiplying the two matrices on the right of Eq. (1.80) results in values of the unknown x's. Matrix A_1 in Eq. (1.79) is obtained from the augmented matrix of Eq. (1.78) by the following procedure.

In order to follow the method described here, relations (1.78) should be considered as a set of equations rather than as two matrices multiplied together. Divide the first row, of the first equation, by a_{11}. Multiply the resulting values of the first row by a_{21} and subtract from the second row; multiply the first row by a_{31} and subtract from the third row. Continuing this will result in a matrix of the form

$$
\begin{bmatrix}
1 & - & - \ldots - & - & 0 & 0 \ldots 0 \\
0 & a'_{22} & - \ldots - & - & 1 & 0 \ldots 0 \\
\vdots & \vdots & & \vdots & \vdots & \\
0 & - & - \ldots - & - & 0 & \\
\vdots & \vdots & & \vdots & \vdots & \\
0 & - & - \ldots a'_{nn} & - & 0 & 1
\end{bmatrix}
\begin{bmatrix}
x_1 \\
x_2 \\
\vdots \\
x_n \\
-b_1 \\
-b_2 \\
\vdots \\
-b_n
\end{bmatrix}
= 0 \qquad (1.81)
$$

where dashes and primes represent constants. Now, divide the second row by a'_{22}. Multiplying the resulting row by $a'_{i2}(i = 3, 4, \ldots, n)$ and subtracting from the third row and higher will give zeros in the second column. Continuing this operation on the third, fourth, and up to the nth rows will result in the matrix

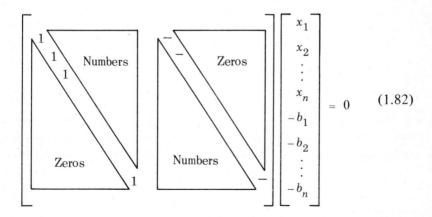

$$
= 0 \qquad (1.82)
$$

Working the above procedure backward from the nth row to the first row, i.e., multiplying the nth row by the element $(n - 1, n)$ and subtracting from the $(n - 2)$ row, and so on, will produce zeros in the nth column. Repeating this process for the $(n - 1)$th row through the first row will result in the matrix

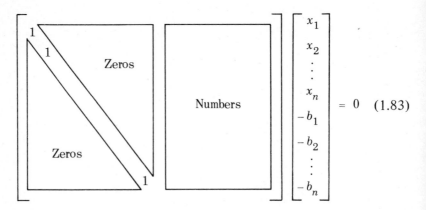

$$= 0 \quad (1.83)$$

Comparing this and Eq. (1.79), we note that the square matrix "numbers" is the inverse of the matrix of the coefficients (or $= [A]^{-1}$).

As mentioned before, the number of multiplications in inverting matrices by the augmented matrix method is estimated at n^3 where n is the order of the matrix to be inverted.

In order to crystallize the above procedures, let us consider the following numerical example.

Example. Find the inverse of the matrix

$$\begin{bmatrix} 2 & 1 & 1 \\ 1 & 3 & 1 \\ 1 & 1 & 4 \end{bmatrix} \quad (1.84)$$

by the augmented matrix method. In numerical evaluation the matrices (B) and (x) can be dropped and the mechanics of arithmetical operations described above can be carried out. The problem starts by writing

$$\begin{bmatrix} 2 & 1 & 1 & 1 & 0 & 0 \\ 1 & 3 & 1 & 0 & 1 & 0 \\ 1 & 1 & 4 & 0 & 0 & 1 \end{bmatrix} \quad (1.85)$$

Dividing the first row by 2, we get

$$\begin{bmatrix} 1 & 0.5 & 0.5 & 0.5 & 0 & 0 \\ 1 & 3 & 1 & 0 & 1 & 0 \\ 1 & 1 & 4 & 0 & 0 & 1 \end{bmatrix} \tag{1.86}$$

Multiplying the first row by −1, adding to the second row, multiplying the first row by −1, and adding to the third row will result in

$$\begin{bmatrix} 1 & 0.5 & 0.5 & 0.5 & 0 & 0 \\ 0 & 2.5 & 0.5 & -0.5 & 1 & 0 \\ 0 & 0.5 & 3.5 & -0.5 & 0 & 1 \end{bmatrix} \tag{1.87}$$

Dividing the second row by 2.5, multiplying the resulting row by −0.5, and adding to the third row, we get

$$\begin{bmatrix} 1 & 0.5 & 0.5 & 0.5 & 0 & 0 \\ 0 & 1 & 0.2 & -0.2 & 0.4 & 0 \\ 0 & 0 & 3.4 & -0.4 & -0.2 & 1 \end{bmatrix} \tag{1.88}$$

Dividing the last row by 3.4, multiplying the resulting row by −0.2, adding to the second row, multiplying the third row by −0.5, adding to the first row, we get

$$\begin{bmatrix} 1 & 0.5 & 0 & 0.5588 & 0.0294 & -0.1470 \\ 0 & 1 & 0 & -0.1764 & 0.4118 & -0.0588 \\ 0 & 0 & 1 & -0.1176 & -0.0588 & 0.2941 \end{bmatrix} \tag{1.89}$$

Multiplying the second row by −0.5 and adding to the first results in

$$\begin{bmatrix} 1 & 0 & 0 & 0.6471 & -0.1765 & -0.1176 \\ 0 & 1 & 0 & -0.1764 & 0.4118 & -0.0588 \\ 0 & 0 & 1 & -0.1176 & -0.0588 & 0.2941 \end{bmatrix} \tag{1.90}$$

From this the inverse matrix of the original matrix will be

$$\begin{bmatrix} 2 & 1 & 1 \\ 1 & 3 & 1 \\ 1 & 1 & 4 \end{bmatrix}^{-1} = \begin{bmatrix} 0.6471 & -0.1765 & -0.1176 \\ -0.1764 & 0.4118 & -0.0588 \\ -0.1176 & -0.0588 & 0.2941 \end{bmatrix} \tag{1.91}$$

FORTRAN Program for Augmented Matrix Inversion Method

The following FORTRAN program is written for matrix inversion by the augmented matrix method discussed above. The data used in the program are the same as the example problem above. The computer program output is given at the end of the program listing.

In order to understand the statements of the program, the reader is referred to [1]. The following paragraphs give a general explanation of the program.

Statement 10 reserves computer memory location for a matrix of 100 by 100. Statements 20 and 30 read the order of the matrix to be inverted, $N = 3$, and the matrix coefficients $S(I, J)$, starting with the first row and proceeding to the second and third rows, as given in Eq. (1.84). Statement 40 and the following two statements set values to be used later. Statements 50 through 80 form the augmented matrix as given by Eq. (1.85). Statements 90 through 220 transform the augmented matrix into the form given by Eq. (1.88) with the last row divided by 3.4. Statements 230 through 290 transform the matrix into the form given by Eq. (1.90). Statements 300 through 330 print out the inverse matrix as given at the end of the program and also by Eq. (1.91).

```
C     MATRIX INVERSION BY AUGMENTED MATRIX METHOD
10    DIMENSION S(100,100)
20    DATA N/3/
```

```
30    DATA ((S(I, J), J=1,3),I=1,3)/ 2.,1.,1.,1.,3.,1.,1.,1.,4./
40    I=1
      NX=N+1
      NY=2*N
50    DO 80 J=NX,NY
60    S(I,J)=1.0
70    I=I+1
80    CONTINUE
90    L=1
100   K=2
110   XM=S(L,L)
120   DO 140 J=L,NY
130   S(L,J)=S(L,J)/XM
140   CONTINUE
150   DO 190 I=K,N
160   X=S(I,L)
170   DO 190 J=L,NY
180   S(I,J)=S(I,J)-S(L,J)*X
190   CONTINUE
200   L=1+1
210   K=K+1
220   IF(L-N) 110,110,230
230   L=N
235   LZ=L-1
240   DO 290 K=1,LZ
250   I=L-K
260   Y=S(I,L)
270   DO 290 J=L,NY
280   S(I,J)=S(I,J)-S(L,J)*Y
290   CONTINUE
300   L=L-1
310   IF(L-1)320,320,235
220   WRITE(2,330) ((S(I,J),J=NX,NY),I=1,N)
330   FORMAT (15H INVERSE MATRIX / (3X1P3E10.3))
340   STOP
350   END
```

Computer solution obtained:

INVERSE MATRIX
    6.471E-01-1.765E-01-1.176E-01
   -1.765E-01 4.118E-01-5.882E-02
   -1.176E-01-5.882E-02 2.941E-01

Inversion of a Matrix by Partitioning

The inversion of large matrices can be accomplished by partitioning the matrix into submatrices. This method reduces the order of the matrices to be inverted, thereby reducing the computer memory storage requirements. Consider the nth order matrix

$$M = \left(\begin{array}{c|c} A & B \\ \hline C & D \end{array}\right) \tag{1.92}$$

partitioned into four submatrices $A, B, C,$ and D where A and D are square matrices of order p and q, respectively; $p + q = n$. The inverse matrix is denoted by

$$M^{-1} = \left(\begin{array}{c|c} W & X \\ \hline Y & Z \end{array}\right) \tag{1.93}$$

where W and Z are square matrices of orders p and q. Using the property of inverse matrices

$$MM^{-1} = \left(\begin{array}{c|c} A & B \\ \hline C & D \end{array}\right)\left(\begin{array}{c|c} W & X \\ \hline Y & Z \end{array}\right) = \left(\begin{array}{cc} I & O \\ O & I \end{array}\right) \tag{1.94}$$

where I is the unit matrix and multiplying Eq. (1.94) we get

$$
\begin{aligned}
AW + BY &= I \\
AX + BZ &= O \\
CW + DY &= O \\
CX + DZ &= I
\end{aligned}
\tag{1.95}
$$

Multiplying the third equation by BD^{-1} and subtracting from the first, we obtain

$$(A - BD^{-1}C)W = I \tag{1.96}$$

or

$$W = (A - BD^{-1}C)^{-1} \tag{1.97}$$

From the third equation we also find that

$$Y = -D^{-1}CW \tag{1.98}$$

From the second and fourth equations we obtain

$$Z = (D - CA^{-1}B)^{-1}$$
$$X = -A^{-1}BZ \tag{1.99}$$

From the above equations, it is seen that the inversion of a matrix of order n reduces to the inversion of two matrices of order p and two of order q and several matrix multiplications.

The equations given above can be altered so that for computation of matrices W, X, Y, Z, only two matrices need be inverted. In terms of the inverse of the square submatrix D, we get

$$W = (A - BD^{-1}C)^{-1}$$
$$X = -WBD^{-1}$$
$$Y = -D^{-1}CW \tag{1.100}$$
$$Z = D^{-1} - D^{-1}CX$$

The above equations can also be set in terms of A^{-1} matrix, as follows:

$$Z = (D - CA^{-1}B)^{-1}$$
$$Y = -ZCA^{-1}$$
$$X = -A^{-1}BZ \tag{1.101}$$
$$W = A^{-1} - A^{-1}BY$$

Note that Eqs. (1.100) and (1.101) involve the inverse of matrices D and A since they were the only square matrices assumed in the partitioning process. In the partitioning method of matrix inversion, since a great number of calculations are involved, it is expedient to check the inverse matrix by multiplying it by the original matrix and obtaining the unit matrix.

Example. Find the inverse of the third-order matrix

$$
M = \left[\begin{array}{cc|c}
2.5 & 3.8 & 9.6 \\
-2.1 & 1.3 & 4.5 \\
\hline
1.0 & 3.2 & 6.7
\end{array}\right]
\tag{1.102}
$$

by partitioning, as indicated.

Partitioned matrices A, B, C, and D of Eq. (1.92) can be written

$$
A = \begin{bmatrix} 2.5 & 3.8 \\ -2.1 & 1.3 \end{bmatrix} \quad
B = \begin{pmatrix} 9.6 \\ 4.5 \end{pmatrix} \quad
C = (1.0,\ 3.2) \quad
D = (6.7)
\tag{1.103}
$$

From this, the inverse of matrix D of Eq. (1.100) will be

$$
D^{-1} = \frac{1}{6.7} = 0.14925
\tag{1.104}
$$

The matrix W of Eq. (1.100) will become

$$
\begin{aligned}
W &= \left(\begin{bmatrix} 2.5 & 3.8 \\ -2.1 & 1.3 \end{bmatrix} - \begin{bmatrix} 1.43280 & 4.58496 \\ 0.67162 & 2.14918 \end{bmatrix} \right)^{-1} \\
&= \begin{pmatrix} 1.06716 & -0.785075 \\ -2.77164 & -0.849254 \end{pmatrix}^{-1} \\
&= \begin{pmatrix} 0.275531 & -0.254709 \\ -0.89923 & -0.34623 \end{pmatrix}
\end{aligned}
\tag{1.105}
$$

Having the above matrix, the second of Eqs. (1.100) can be used to calculate X. This results in

$$X = \begin{pmatrix} 0.275531 & -0.254709 \\ -0.89923 & -0.34623 \end{pmatrix} \begin{pmatrix} 9.6 \\ 4.5 \end{pmatrix} (0.14925)$$

$$= \begin{pmatrix} -0.223718 \\ 1.52099 \end{pmatrix}$$

Using the previously calculated values, the third and fourth relations of Eq. (1.100) will result in values of Y and Z as follows

$$Y = (0.388359 \quad 0.20338)$$

$$Z = -0.543799$$

Setting the values W, X, Y, and Z in Eq. (1.93) of the inverse matrix, we get

$$M^{-1} = \begin{pmatrix} 0.275531 & -0.254709 & -0.223718 \\ -0.89923 & -0.34623 & 1.52099 \\ 0.388359 & 0.20338 & -0.543799 \end{pmatrix} \tag{1.106}$$

which is the required answer to the problem.

Gauss-Seidel Method of Solving Simultaneous Equations

A system of linear equations is called diagonal when the coefficient of the diagonal element dominates, that is, the absolute value of this coefficient is larger than the sum of the absolute values of the rest of the coefficients on that row. The simultaneous equations arising in many physical systems satisfy this condition. In these cases the iterative method of Gauss-Seidel can be applied for the solution of the unknowns.

The mathematical derivation of the convergence criteria of the Gauss-Seidel iteration method for solving sets of simultaneous equations is complicated as the criterion of convergence is related to eigenvalues (see Chap. 2) of a matrix derived from the matrix of the coefficients, matrix A in the following discussion [7]. It will be seen in the succeeding chapter that the computation of eigenvalues for high-order matrices is a difficult numerical problem. For this reason, in practical cases, the convergence criterion of Gauss-Seidel method can be applied only in the case of sets of simultaneous equations consisting of very few equations— 2 or 3.

Given the set of simultaneous equations

$$a_{11}x_1 + a_{12}x_2 + \cdots + a_{1n}x_n = b_1$$

$$a_{21}x_1 + a_{22}x_2 + \cdots + a_{2n}x_n = b_2$$

$$\vdots$$

$$a_{n1}x_1 + a_{n2}x_2 + \cdots + a_{nn}x_n = b_n$$

(1.107)

or, in matrix notation,

$$Ax = B \tag{1.108}$$

Equations (1.107) are solved for the diagonal unknowns

$$x_1 = \frac{b_1}{a_{11}} - \left(\frac{a_{12}}{a_{11}}x_2 + \frac{a_{13}}{a_{11}}x_3 + \cdots + \frac{a_{1n}}{a_{11}}x_n \right)$$

$$x_2 = \frac{b_2}{a_{22}} - \left(\frac{a_{21}}{a_{22}}x_1 + \frac{a_{23}}{a_{22}}x_3 + \cdots + \frac{a_{2n}}{a_{22}}x_n \right)$$

$$\vdots$$

$$x_n = \frac{b_n}{a_{nn}} - \left(\frac{a_{n1}}{a_{nn}}x_1 + \frac{a_{n2}}{a_{nn}}x_2 + \cdots + \frac{a_{n,n-1}}{a_{nn}}x_{n-1} \right)$$

(1.109)

and in matrix notation

$$x = B' - Cx \tag{1.110}$$

where C is a matrix with diagonal elements of zero. Gauss-Seidel iterations start by assuming that $x_2 = x_3 = \cdots = x_n = 0$ and solving the first of Eqs. (1.109) for x_1 using this value of x_1 with $x_3 = x_4 = \cdots = x_n = 0$, solving the second of Eqs. (1.109) for x_2, and so on. After obtaining the set of values for x_1 through x_n, the second iteration starts by using these values in Eqs. (1.109). In most practical cases solution is terminated where the values of the unknowns from one iteration to the next do not change within a specified number of significant figures.

Example. Solve the set of equations

$$25x_1 + 2x_2 + x_3 = 69$$
$$2x_1 + 10x_2 + x_3 = 63$$
$$x_1 + x_2 + 4x_3 = 43$$

$$(1.111)$$

The solution starts by putting the above equation in the form (1.109) and carrying out the above outlined procedure. The values of x at the end of the first iteration are

$$x_1 = 2.76 \quad x_2 = 5.748 \quad x_3 = 8.623$$

The solution will converge after five iterations to the values

$$x_1 = 2 \quad x_2 = 5 \quad x_3 = 9$$

Computer Program for the Above Solution

The following digital computer program in BASIC programming language was written for the Gauss-Seidel iteration method. Although most of the computer instructions are self-explanatory, the reader is referred to the BASIC programming language manual [2] for further explanation of instructions.

```
100   DIM A(10,10),C(10,10),B(10),X(10)
110   READ N
120   PRINT "COEFFICIENT"
130   FOR I=1 TO N
140   FOR J=1 TO N
150   READ A(I,J)
160   PRINT A(I,J);
170   NEXT J
180   PRINT
190   NEXT I
200   FOR I=1 TO N
210   FOR J=1 TO N
220   LET C(I,J)=A(I,J)/A(I,I)
230   NEXT J
```

```
240   LET  C(I,I)=0
250   NEXT  I
260   PRINT "RIGHT SIDE"
270   FOR I=1 TO N
280   READ B(I)
290   PRINT B(I);
300   NEXT I
310   PRINT
320   FOR I=1 TO N
330   LET X(I)=0
340   NEXT I
350   LET L=1
360   PRINT "ITERATION     VALUES OF X(1) THRU X(N)"
370   FOR I=1 TO N
380   LET D=0
390   FOR J=1 TO N
400   LET D=D+C (I,J)*X(J)
410   NEXT J
420   LET X(I)=(B(I)/A(I,I))-D
430   NEXT I
440   PRINT L;
450   FOR I=1 TO N
460   PRINT X(I);
470   NEXT I
480   PRINT
490   IF L=10 THEN 570
500   LET L=L+1
510   GO TO 370
520   DATA 3
530   DATA 25,2,1
540   DATA 2,10,1
550   DATA 1,1,4
560   DATA 69,63,43
570   END
```

Statement 100 of the program reserves computer memory spaces for the coefficients of Eq. (1.108). The matrix C of Eq. (1.110) is also shown in the dimension statement 100. Statement 110 reads the number of the simultaneous equations. Statements 130 through 190 read the coefficients of the unknowns of Eqs. (1.107). Statements 200 through 250 compute the terms of the matrix C of Eq. (1.110). Note that the diagonal elements of this matrix are set to zero by statement 240. Statements 270 through 300 read the elements of the right side of Eq.

(1.107).   Statements 370 to 430 compute the values of x's at each iteration according to Eq. (1.110).   Statements 450 through 470 print out the answers.   Statement 490 limits the number of iterations to 10. The output of the program for 5 iterations of the above example problem is as follows:

```
COEFFICIENT
25   2    1
2    10   1
1    1    4
RIGHT SIDE
69   63   43
ITERATION      VALUES OF X[1] THRU X[N]
1    2.76      5.748      8.623
2    1.95524   5.04665    8.99953
3    1.99629   5.00079    9.00073
4    1.99991   4.99995    9.00004
5    2.        5.         9.
```

In this chapter we have discussed the introductory material on matrices and linear simultaneous equations.   However, we have not treated the question of round-off errors in the computation and the stability of solutions.   There are a number of texts which discuss these problems from a mathematical point of view [4, 7].   In most engineering applications, however, these questions are studied by an examination of the answers obtained for a variety of input parameter conditions.   For example, the simplest method of examining the correctness of the inverse matrix A^{-1} is to multiply it by A and obtain the unit matrix.   The question of the stability of solution, for example, for matrix inversion, can be answered by perturbing one or some of the matrix elements slightly and examining its effect on the solution.

PROBLEMS

1. Given the two equations

$$ax + 3y = 0$$

$$2x + (a - 1)y = 0$$

for what value of a do these equations have a nonzero solution?

Find the value of the ratio x/y for each value of a.

2. Expand the determinant

$$D = \begin{vmatrix} x & y & z \\ z & x & y \\ y & z & x \end{vmatrix}$$

3. Evaluate the determinant

$$D = \begin{vmatrix} 10 & 1 & 1 \\ 20 & 4 & 6 \\ 30 & 12 & 9 \end{vmatrix}$$

4. Verify that $x = 2$ is a root of the equation,

$$\begin{vmatrix} (10 - x) & -6 & 2 \\ -6 & (9 - x) & -4 \\ 2 & -4 & (5 - x) \end{vmatrix} = 0$$

5. Evaluate the determinant

$$\begin{vmatrix} 15 & 1 & 2 & -3 \\ 5 & 6 & 4 & 4 \\ -10 & -3 & 2 & 1 \\ -5 & 3 & 4 & 0 \end{vmatrix}$$

by the method of Chió.   *Answer:* −1820

6. Evaluate the determinant

$$\begin{vmatrix} 7 & 2 & 4 \\ 3 & -4 & 5 \\ 1 & 3 & -2 \end{vmatrix}$$

by following the logic of the FORTRAN program given for determinant evaluation.   *Answer:* 25.

7. Show that

$$D = \begin{vmatrix} 2\cos\theta & 1 & 0 \\ 1 & 2\cos\theta & 1 \\ 0 & 1 & 2\cos\theta \end{vmatrix} = \frac{\sin 4\theta}{\sin\theta}$$

provided that θ is not a multiple of π.

8. Solve the equations

$$9x - 17y + 21z = 66$$
$$13x + 8y - 7z = 32$$
$$7x - 3y - 9z = -37$$

Answer: $x = 2.818$, $y = 3.821$, $z = 5.029$.

9. Find the values of a for which the equations

$$(1 - a)x - 6y + 4z = 0$$
$$-6x + (2 - a)y - 2z = 0$$
$$4x - 2y - (3 + a)z = 0$$

can be satisfied by values of x, y, z, not all zero. For each value of a, find the ratios $x:y:z$.

10. Solve the set of equations for x_1, x_2, and x_3 by the method of elimination of Gauss

$$7x_1 + 2x_2 + 4x_3 = 9$$
$$3x_1 - 4x_2 + 5x_3 = -14$$
$$x_1 + 3x_2 - 2x_3 = 12$$

Answer: $x_1 = 1$, $x_2 = 3$, $x_3 = -1$.

11. Solve Prob. 10 by triangulation method.
12. Show that the equations

$$x + y + cz = m$$

$$2x + 3y + 4z = 0$$

$$3x + 4y + 5z = 1$$

have a unique solution if $c \neq 1$ and find it. Discuss the case $c = 1$.

13. Prove that

$$\begin{vmatrix} a & b & c \\ c & a & b \\ b & c & a \end{vmatrix} \begin{vmatrix} x & y & z \\ z & x & y \\ y & z & x \end{vmatrix} = \begin{vmatrix} p & q & r \\ r & p & q \\ q & r & p \end{vmatrix}$$

where $p = ax + by + cz$, $q = az + bx + cy$, $r = ay + bz + cx$.

14. Verify that

$$\begin{vmatrix} 1 & -5 & 5 & 3 \\ 2 & -3 & 3 & 5 \\ 2 & -3 & 2 & 9 \\ 1 & 2 & -2 & 2 \end{vmatrix} = 0$$

15. If

$$A = \begin{bmatrix} 2 & 3 & 1 \\ 0 & 2 & -1 \end{bmatrix} \qquad B = \begin{bmatrix} 1 & -2 \\ 3 & 0 \\ -1 & 2 \end{bmatrix}$$

find AB and BA.

16. If

$$A = \begin{bmatrix} 2 & 3 \\ 0 & 0 \end{bmatrix} \qquad B = \begin{bmatrix} 3 & 0 \\ -2 & 0 \end{bmatrix}$$

show that $AB = 0$. This shows that from $AB = 0$ we cannot deduce that $A = 0$ or $B = 0$.

17. If

$$A = \begin{bmatrix} 0 & 0 \\ 2 & 0 \end{bmatrix}$$

show that $A^2 = 0$, and hence that we cannot deduce from $A^2 = 0$ that $A = 0$.

18. Show that the commutative law holds for the product of two matrices of the form

$$A = \begin{bmatrix} a & b \\ -b & a \end{bmatrix} \quad \text{and} \quad B = \begin{bmatrix} x & y \\ -y & x \end{bmatrix}$$

19. If A and B are square matrices of the same order, show that, unless $AB = BA$,

   a. $(A + B)(A - B) \neq A^2 - B^2$

   b. $(A + B)^2 \neq A^2 + 2AB + B^2$

   c. $(A - 2B)(A - 3B) \neq A^2 - 5AB + 6B^2$

20. If

$$A = \begin{bmatrix} 1 & 2 \\ 3 & 4 \end{bmatrix}$$

show that the products AA' and $A'A$ are symmetric but not equal.

21. If

$$A = \begin{bmatrix} 1 & 0 & 1 \\ 8 & 1 & 2 \\ 0 & -1 & 6 \end{bmatrix}$$

show that A is singular. Construct adj A.

22. If

$$A = \begin{bmatrix} 2 & 2 & 2 \\ 2 & 5 & 5 \\ 2 & 5 & 11 \end{bmatrix} \quad \text{show that} \quad A^{-1} = \frac{1}{6}\begin{bmatrix} 5 & -2 & 0 \\ -2 & 3 & -1 \\ 0 & -1 & 1 \end{bmatrix}$$

23. If

$$A = \begin{bmatrix} 1 & 1 & 1 \\ 1 & a & a^2 \\ 1 & a^2 & a \end{bmatrix} \quad \text{then} \quad A^{-1} = \frac{1}{3}\begin{bmatrix} 1 & 1 & 1 \\ 1 & a^2 & a \\ 1 & a & a^2 \end{bmatrix}$$

where $a = \exp(2\pi i/3)$, $i = (-1)^{1/2}$.

24. If A is a square matrix of order n and if x is a scalar, show that

$$|Ax| = |A|x^n \quad \text{and that} \quad \text{adj}(Ax) = (\text{adj } A)x^{n-1}$$

25. Obtain the inverse of the matrix

$$[A] = \begin{bmatrix} 6 & -2 & 3 \\ 1 & -4 & -2 \\ 3 & 1 & 3 \end{bmatrix}$$

by the augmented matrix method. Check your solution by multiplying the inverse matrix $[A]^{-1}$ by matrix $[A]$.

26. Follow the logic of the FORTRAN program given for the augmented matrix method in inverting the matrix of Prob. 25.

27. Verify Eqs. (1.100) and (1.101), which were used to obtain inverse matrices by matrix partitioning.

28. Obtain the inverse of the M matrix

$$M = \begin{bmatrix} 2.5 & 3.8 & 9.6 \\ -2.1 & 1.3 & 4.5 \\ 1.0 & 3.2 & 6.7 \end{bmatrix}$$

by partitioning the matrix and using Eqs. (1.101).

29. Obtain the inverse of matrix of Prob. 25 by the matrix partitioning method.

30. Solve the set of simultaneous equations

$$25x_1 + 2x_2 + x_3 = 69$$
$$2x_1 + 10x_2 + x_3 = 63$$
$$x_1 + x_2 + 4x_3 = 43$$

by the matrix inversion method.

31. Solve the set of equations

$$2x_1 - 3x_2 + x_3 = -4$$
$$x_1 + x_2 - 2x_3 = -7$$
$$-x_1 + x_2 + x_3 = 9$$

by the matrix inversion method.

32. Solve the set of simultaneous equations

$$4x_1 + x_2 - x_3 = 17$$
$$x_1 + 6x_2 - 2x_3 = 14$$
$$x_1 + x_2 - 5x_3 = 1$$

by the Gauss-Seidel method.

33. Solve the set of equations

$$3x_1 + x_2 + x_3 = 4$$
$$x_1 + 4x_2 - x_3 = -5$$
$$x_1 + x_2 - 6x_3 = -12$$

by the Gauss-Seidel method.

REFERENCES

1. McCracken, D. D.: "A Guide to FORTRAN IV Programming," John Wiley & Sons, Inc., New York, 1965.
2. BASIC Language, reference manual, General Electric Company, May, 1966.
3. Pipes, L. A.: "Matrix Methods for Engineering," Prentice-Hall, Inc., Englewood Cliffs, N. J., 1963.
4. Faddeev, D. K., and V. N. Faddeeva: "Computational Methods of Linear Algebra," W. H. Freeman & Co., Publishers, San Francisco, 1963.
5. Kunz, Kaiser S.: "Numerical Analysis," McGraw-Hill Book Company, New York, 1957.
6. Salvadori, Mario G., and Melvin L. Baron: "Numerical Methods in Engineering," Prentice-Hall, Inc., Englewood Cliffs, N. J., 1964.
7. Todd, John (ed.): "Survey of Numerical Analysis," McGraw-Hill Book Company, New York, 1962.

2 EIGENVALUES, EIGENVECTORS, AND QUADRATIC FORMS

The study of many physical systems of practical importance, such as the investigation of the elastic vibrations of a bridge or any other solid structure, the flutter of an airplane wing, the transient oscillations of an electric network, or the buckling of an elastic structure, leads to the solution of what is known in the mathematical literature as an eigenvalue problem. A general mathematical discussion of the eigenvalue problem and its ramifications is given in this chapter. This class of problems is also referred to as characteristic value problems. The terms "eigen" and "characteristic" are used interchangeably.

THE EIGENVECTOR (CHARACTERISTIC VECTOR) EQUATION

Let us assume that we are given an nth order square matrix A and an nth-order column vector X. Then A and X are conformable matrices and can be multiplied in the order AX to yield another vector Y, that is,

$$AX = Y \tag{2.1}$$

In general, the vector Y will not have the same direction or magnitude as the vector X. Let us ask the question of whether or not it may happen that the vector Y has the same direction as the vector X. If this is the case, $Y = \lambda X$ where λ is a multiplicative scalar. If we place $Y = \lambda X$ in Eq. (2.1), we obtain

$$AX = \lambda X \tag{2.2}$$

Equation (2.2), usually called the matrix *eigenvector equation*, arises in many branches of applied mathematics. The main problem is to find the values of λ and the corresponding vector X for which Eq. (2.2) holds. Before considering the general case, let us consider the case of a second-order square matrix A given by

$$A = \begin{bmatrix} a & b \\ c & d \end{bmatrix} \quad \text{and} \quad X = \begin{bmatrix} x_1 \\ x_2 \end{bmatrix} \tag{2.3}$$

If we expand the eigenvector equation (2.2) in this case we obtain

$$\begin{bmatrix} a & b \\ c & d \end{bmatrix} \begin{bmatrix} x_1 \\ x_2 \end{bmatrix} = \lambda \begin{bmatrix} x_1 \\ x_2 \end{bmatrix} \tag{2.4}$$

If we multiply the matrix A by the column vector X as indicated in Eq. (2.4), we obtain the two scalar equations

$$\begin{aligned} ax_1 + bx_2 &= \lambda x_1 \\ cx_1 + dx_2 &= \lambda x_2 \end{aligned} \tag{2.5}$$

Equations (2.5) can be rewritten in the form

$$\begin{aligned} (a - \lambda)x_1 + bx_2 &= 0 \\ cx_1 + (d - \lambda)x_2 &= 0 \end{aligned} \tag{2.6}$$

Equations (2.6) are a set of homogeneous equations of the type discussed in Chap. 1. In order for these equations to have nontrivial solutions for x_1 and x_2, it is necessary that the determinant of the coefficients $p(\lambda)$ vanish, that is,

$$
p(\lambda) = \begin{vmatrix} a - \lambda & b \\ c & d - \lambda \end{vmatrix} \tag{2.7}
$$

$$
= \det |A - \lambda I|
$$

$$
= 0
$$

The matrix $K(\lambda) = [A - \lambda I]$ is called the *characteristic matrix* of the matrix A; the polynomial $p(\lambda) = \det(A - \lambda I)$ is called the *characteristic polynomial.* Equation (2.7) or $p(\lambda) = 0$ is the *characteristic equation* of the matrix A.

THE EIGENVALUES OF A

The characteristic equation (2.7) of the matrix A has the form

$$
p(\lambda) = \lambda^2 - \lambda(a + d) + (ad - bc) = 0 \tag{2.8}
$$

The two roots λ_1 and λ_2 of this quadratic equation are the *eigenvalues* of the second-order matrix A given by Eq. (2.3).

If λ_1 and λ_2 are the roots of Eq. (2.8), we can write the polynomial $p(\lambda)$ in the following factored form:

$$
p(\lambda) = (\lambda - \lambda_1)(\lambda - \lambda_2) = \lambda^2 - \lambda(\lambda_1 + \lambda_2) + \lambda_1\lambda_2 = 0 \tag{2.9}
$$

If we now compare Eqs. (2.8) and (2.9), we see that the coefficient of the $-\lambda$ term $(a + d)$ of Eq. (2.8) must be equal to the coefficient of the $-\lambda$ term of Eq. (2.9), and we must therefore have

$$
\lambda_1 + \lambda_2 = (a + d) = \text{Tr } A = \text{Trace of } A \tag{2.10}
$$

The constant term of Eq. (2.8) must correspond to the constant term of Eq. (2.9); therefore, we must have

$$
\lambda_1\lambda_2 = (ad - bc) = \det A = \text{determinant of } A \tag{2.11}
$$

The important fact that the sum of the eigenvalues of a matrix A equals the sum of the principal diagonal elements of A or the trace of A and that the product of the eigenvalues of A equals the determinant of A is true not only for second-order matrices, but is true also in the general case of a matrix of order n. This fact will be established later. The two eigenvalues can be obtained by solving the quadratic equation (2.8) (the characteristic equation of A).

THE EIGENVECTORS OF A

The eigenvalues of A are the numbers λ_1 and λ_2. In order to obtain the eigenvectors of A that correspond to these eigenvalues, we return to the simultaneous equations (2.6). From these equations, we may obtain the ratio x_2/x_1 in the form

$$\frac{x_2}{x_1} = \frac{\lambda - a}{b} = \frac{c}{\lambda - d} \tag{2.12}$$

If we place $\lambda = \lambda_1$ in Eq. (2.12) we obtain the ratio

$$r_1 = \frac{\lambda_1 - a}{b} \tag{2.13}$$

We thus obtain $x_2 = r_1 x_1$. This fixes the ratio of x_2 to x_1 and therefore the direction of a certain vector. If we place $\lambda = \lambda_2$ in Eq. (2.12) we obtain

$$r_2 = \frac{\lambda_2 - a}{b} \tag{2.14}$$

We thus obtain $x_2 = r_2 x_1$, and therefore another direction of a vector. We can write X_1 for a vector of undetermined length having the direction r_1, and X_2 for a vector of undetermined length having the direction r_2, and obtain

$$X_1 = \begin{vmatrix} x_1 \\ r_1 x_1 \end{vmatrix} \qquad X_2 = \begin{vmatrix} x_1 \\ r_2 x_1 \end{vmatrix} \tag{2.15}$$

The vector X_1 is then said to belong to the eigenvalue λ_1 and the vector X_2 to the eigenvalue λ_2. Since the quantity x_1 is not specified, only the directions of these vectors are fixed. If we choose x_1 to be given by

$$x_1 = \frac{1}{\left(1 + r_1^2\right)^{1/2}} \quad \text{and} \quad x_1 = \frac{1}{\left(1 + r_2^2\right)^{1/2}} \tag{2.16}$$

in X_1 and X_2, respectively, the vectors will then be normalized to have unit length.

It is seen from the above discussion that the eigenvector equation gives eigenvalues and eigendirections. In most discussions involving eigenvalues the vectors are normalized to unit length and in such case they are uniquely defined.

THE EIGENVALUES AND EIGENVECTORS
OF AN nTH-ORDER SQUARE MATRIX

If we expand the eigenvector equation (2.2) in the general case in which the matrix A is a square matrix of the nth order, we have

$$
\begin{aligned}
a_{11}x_1 + a_{12}x_2 + \cdots + a_{1n}x_n &= \lambda x_1 \\
a_{21}x_1 + a_{22}x_2 + \cdots + a_{2n}x_n &= \lambda x_2 \\
&\cdots\cdots\cdots\cdots\cdots\cdots\cdots \\
a_{n1}x_1 + a_{n2}x_2 + \cdots + a_{nn}x_n &= \lambda x_n
\end{aligned}
\tag{2.17}
$$

This set of equations can be written in the form

$$
\begin{aligned}
(a_{11} - \lambda)x_1 + a_{12}x_2 + \cdots + a_{1n}x_n &= 0 \\
a_{21}x_1 + (a_{22} - \lambda)x_2 + \cdots + a_{2n}x_n &= 0 \\
&\vdots \\
a_{n1}x_1 + a_{n2}x_2 + \cdots + (a_{nn} - \lambda)x_n &= 0
\end{aligned}
\tag{2.18}
$$

In matrix notation, Eq. (2.18) can be written in the compact form

$$[A - \lambda I](X) = 0 \tag{2.19}$$

where I is the unit matrix of order n. The homogeneous equations (2.18) have a nonzero solution only in the case where the determinant $p(\lambda)$ of the coefficients vanishes. We therefore have

$$p(\lambda) = \det [A - \lambda I] = 0 \tag{2.20}$$

In matrix algebra, the matrix

$$K(\lambda) = [A - \lambda I] \tag{2.21}$$

is called the *characteristic matrix* of the matrix A. The determinant of the characteristic matrix $p(\lambda)$ is the *characteristic polynomial* of A. This determinant is generally of order n and therefore is a polynomial of the nth degree in λ. The equation $p(\lambda) = 0$ is the *characteristic equation* of the matrix A.

FUNDAMENTAL PROPERTIES OF THE CHARACTERISTIC POLYNOMIAL OF A

If A is an nth order matrix the characteristic polynomial can also be written in the form $p(\lambda) = \det[\lambda I - A] = 0$. In this form it results in the characteristic equation with λ^n coefficient of unity,

$$p(\lambda) = \lambda^n + a_1 \lambda^{n-1} + a_2 \lambda^{n-2} + \cdots + a_{n-1} \lambda + a_n = 0 \tag{2.22}$$

Setting $\lambda = 0$ in Eq. (2.22) will result in

$$p(0) = a_n \tag{2.23}$$

However, since $p(\lambda) = \det [\lambda I - A]$, we have

$$p(0) = \det [-A] = (-1)^n \det [A] \tag{2.24}$$

If we substitute Eq. (2.23) into Eq. (2.24), we obtain

$$a_n = (-1)^n \det A \tag{2.25}$$

If we assume that the matrix A has n distinct eigenvalues $(\lambda_1, \lambda_2, \ldots, \lambda_n)$, then since the eigenvalues are the roots of the characteristic equation, we can write Eq. (2.22) in the factored form

$$p(\lambda) = (\lambda - \lambda_1)(\lambda - \lambda_2) \ldots (\lambda - \lambda_n) \tag{2.26}$$

If we place $\lambda = 0$ in Eq. (2.26), the result is

$$p(0) = (-1)^n \lambda_1 \lambda_2 \lambda_3 \ldots \lambda_n \tag{2.27}$$

If we compare Eqs. (2.24) and (2.27), we have

$$\det [A] = \lambda_1 \lambda_2 \ldots \lambda_n \tag{2.28}$$

Therefore the determinant of the nth order square matrix A is equal to the product of the eigenvalues of A. It follows that if $\det [A] = 0$ so that A is a singular matrix, then at least one of its eigenvalues is zero.

If the factored form of $p(\lambda)$ of Eq. (2.26) is multiplied out, it will be found that the coefficient of λ^{n-1} which is a_1 in Eq. (2.22) is given by

$$\lambda_1 + \lambda_2 + \lambda_3 + \cdots + \lambda_n = -a_1 \tag{2.29}$$

If, on the other hand, the determinant, $\det [A - \lambda I]$, is expanded to determine the coefficient of a_1 or λ^{n-1} in the polynomial $p(\lambda)$, the result is

$$a_1 = -(a_{11} + a_{22} + a_{33} + \cdots + a_{nn}) = -\text{Trace } A \tag{2.30}$$

since by definition the trace of the matrix A, $\text{Tr}A$, is the sum of the principal diagonal terms of A. If Eqs. (2.29) and (2.30) are compared, it is seen that we have the important relation

$$\text{Trace } A = \lambda_1 + \lambda_2 + \lambda_3 + \cdots + \lambda_n \tag{2.31}$$

The above equations can be used primarily as a check on computation of the eigenvalues. The nth-order polynomial of Eq. (2.22) can be solved for characteristic values by the methods outlined in Chap. 3.

NEWTON'S FORMULAS FOR THE COEFFICIENTS OF THE CHARACTERISTIC POLYNOMIAL

We have seen that the coefficient a_n of the characteristic polynomial $p(\lambda)$ is given by the relation $a_n = \det A$ and that the coefficient

$a_1 = -$Trace A. The mathematician M. Bôcher has given a set of equations for the coefficients of the characteristic polynomial $p(\lambda)$.*

We define the numbers $S_1, S_2, S_3, \ldots, S_n$ by the equations

$$S_1 = \text{Tr}(A) \quad S_2 = \text{Tr}(A^2) \quad S_3 = \text{Tr}(A^3) \ldots S_n = \text{Tr}(A^n) \quad (2.32)$$

so that S_k is the trace of the kth power of the given nth-order square matrix A. Bôcher has shown by a long algebraic argument based on Newton's formulas that the coefficients $a_1, a_2, \ldots, a_n$ of the characteristic polynomial $p(\lambda)$ of A can be computed successively by the following recurrence formulas:

$$a_1 = -S_1$$

$$a_2 = -\frac{a_1 S_1 + S_2}{2}$$

$$a_3 = -\frac{a_2 S_1 + a_1 S_2 + S_3}{3} \qquad (2.33)$$

$$\ldots\ldots\ldots\ldots\ldots\ldots\ldots\ldots\ldots\ldots\ldots\ldots\ldots\ldots$$

$$a_n = -\frac{a_{n-1} S_1 + a_{n-2} S_2 + \cdots + a_1 S_{n-1} + S_n}{n}$$

As a numerical example of the use of these formulas, let it be required to determine the characteristic polynomial and therefore the characteristic equation of the matrix

$$A = \begin{bmatrix} 2 & -1 & 0 \\ 9 & 4 & 6 \\ -8 & 0 & -3 \end{bmatrix} \qquad (2.34)$$

We first compute A^2 and A^3 and obtain

*M. Bôcher, "Introduction to Higher Algebra," pp. 243-244, Dover Publications, Inc., New York, 1964.

$$A^2 = \begin{bmatrix} -5 & -6 & -6 \\ 6 & 7 & 6 \\ 8 & 8 & 9 \end{bmatrix} \quad A^3 = \begin{bmatrix} -16 & -19 & -18 \\ 27 & 22 & 24 \\ 16 & 24 & 21 \end{bmatrix} \tag{2.35}$$

We then compute the traces

$$S_1 = \text{Tr } A = 3 \quad S_2 = \text{Tr } (A^2) = 11 \quad S_3 = \text{Tr } (A^3) = 27 \tag{2.36}$$

By the use of Eq. (2.33) we then obtain

$$a_1 = -S_1 = -3$$

$$a_2 = -\frac{a_1 S_1 + S_2}{2} = -1 \tag{2.37}$$

$$a_3 = -\frac{a_2 S_1 + a_1 S_2 + S_3}{3} = 3$$

The characteristic equation of the matrix (2.34) is, therefore,

$$\lambda^3 - 3\lambda^2 - \lambda + 3 = 0 \tag{2.38}$$

Equation (2.25) in this case gives $\det A = -3$.

It should be noted that although Bôcher's formulas represent an elegant mathematical relation, in practical cases they are not used in computation of the coefficients of the characteristic polynomial of high order (3d and higher) matrices. The reason for this is that raising matrices to various powers, in addition to requiring a great number of calculations, in many cases results in loss of accuracy as the elements of the new matrices require more significant figures. The following numerical method, due to A. N. Krylov, can be used for the determination of the coefficients of the characteristic equation.

THE METHOD OF A. N. KRYLOV FOR DETERMINATION OF THE CHARACTERISTIC EQUATION

Consider the matrix

$$
A = \begin{vmatrix}
a_{11} & a_{12} & a_{13} \cdots a_{1n} \\
a_{21} & a_{22} & a_{23} \cdots a_{2n} \\
\vdots & & \\
a_{n1} & a_{n2} & a_{n3} \cdots a_{nn}
\end{vmatrix}
\tag{2.39}
$$

whose characteristic equation is desired. As discussed before, the characteristic equation results from setting the determinant

$$
D(\lambda) = \begin{vmatrix}
a_{11} - \lambda & a_{12} & \cdots a_{1n} \\
a_{21} & a_{22} - \lambda & \cdots a_{2n} \\
\vdots & & \\
a_{n1} & a_{n2} & \cdots a_{nn} - \lambda
\end{vmatrix}
\tag{2.40}
$$

equal to zero. The method of A. N. Krylov consists of reducing the determinant of Eq. (2.40) to the form

$$
D_n(\lambda) = \begin{vmatrix}
b_{11} - \lambda & b_{12} \cdots b_{1n} \\
b_{21} - \lambda^2 & b_{22} \cdots b_{2n} \\
\vdots & & \\
b_{n1} - \lambda^n & b_{n2} \cdots b_{nn}
\end{vmatrix} = 0
\tag{2.41}
$$

where the characteristic value λ appears only in the first column. Expanding the above determinant in terms of its first column will give the characteristic equation in powers of λ. Note that the form of Eq. (2.41) is such that the characteristic polynomial can more readily be obtained than from the expansion of the determinant of Eq. (2.40).

To derive the above relation we start by noting that Eq. (2.40) is the necessary and sufficient condition for the system of homogeneous equations

$$
\begin{aligned}
\lambda x_1 &= a_{11}x_1 + a_{12}x_2 + \cdots + a_{1n}x_n \\
\lambda x_2 &= a_{21}x_1 + a_{22}x_2 + \cdots + a_{2n}x_n \\
&\ \vdots \\
\lambda x_n &= a_{n1}x_1 + a_{n2}x_2 + \cdots + a_{nn}x_n
\end{aligned}
\tag{2.42}
$$

to have a solution $x_1, x_2, x_3, \ldots, x_n$ different from zero. Multiplying the first equation by λ and replacing the right side values of $\lambda x_1, \lambda x_2, \lambda x_3, \ldots$ of the resulting equation by their equivalents from Eq. (2.42), we get

$$\lambda^2 x_1 = b_{21} x_1 + b_{22} x_2 + \cdots + b_{2n} x_n \tag{2.43}$$

where

$$b_{2k} = \sum_{s=1}^{n} a_{1s} a_{sk} \qquad k = 1, \ldots, n$$

Multiplying Eq. (2.43) again by λ and replacing the right side of the resulting equation by equivalent values from Eq. (2.42), we get

$$\lambda^3 x_1 = b_{31} x_1 + b_{32} x_2 + \cdots + b_{3n} x_n \tag{2.44}$$

Repeating the above process will result in the set of equations

$$\begin{aligned}
\lambda x_1 &= b_{11} x_1 + b_{12} x_2 + \cdots + b_{1n} x_n \\
\lambda^2 x_1 &= b_{21} x_1 + b_{22} x_2 + \cdots + b_{2n} x_n \\
&\ \ \vdots \\
\lambda^n x_1 &= b_{n1} x_1 + b_{n2} x_2 + \cdots + b_{nn} x_n
\end{aligned} \tag{2.45}$$

where

$$b_{1k} = a_{1k}$$

$$b_{ik} = \sum_{s=1}^{n} b_{i-1,s} a_{sk} \qquad \begin{array}{l} i = 2, \ldots, n \\ k = 1, \ldots, n \end{array}$$

The set of equations (2.45) will have a nonzero solution in x's only if the condition on the determinant of Eq. (2.41) is satisfied.

Expanding the determinant of Eq. (2.41) about the first column and denoting the cofactors of element (i, j) by $C_{i,j}$, we get

$$C_{n,1}\lambda^n + C_{n-1,1}\lambda^{n-1} + C_{n-2}\lambda^{n-2} + \cdots$$

$$+ C_{1,1}\lambda - \sum_{i=1}^{n} b_{i1}C_{i1} = 0 \tag{2.46}$$

Dividing through by the coefficient of the λ^n term, we get

$$\lambda^n + P_{n-1}\lambda^{n-1} + P_{n-2}\lambda^{n-2} + \cdots + P_1\lambda + P_0 = 0 \tag{2.47}$$

where

$$p_j = \frac{C_{j,1}}{C_{n,1}} \qquad j = 1, \ldots, n-1$$

$$p_0 = -\sum_{i=1}^{n} b_{i1}\frac{C_{i1}}{C_{n,1}}$$

In Eq. (2.47) the p's are the coefficients of the characteristic equation of the matrix A of Eq. (2.39). Having obtained this equation the roots can be evaluated by root finding methods of Chap. 3. It should be noted that in many physical problems a great deal of information can be extracted from the characteristic equation without actually solving for the roots.

Example. It is required to find the characteristic equation of the determinant

$$A = \begin{vmatrix} 2 & 4 & -6 \\ 4 & 2 & -6 \\ -6 & -6 & -15 \end{vmatrix}$$

From Eqs. (2.45) the values of the coefficients b_{ij} can be calculated

$$b_{11} = a_{11} = 2 \qquad b_{12} = a_{12} = 4 \qquad b_{13} = a_{13} = -6$$

and

$$b_{21} = \sum_{s=1}^{3} b_{1s} a_{s1} = b_{11} a_{11} + b_{12} a_{21} + b_{13} a_{31} = 56$$

$$b_{22} = \sum_{s=1}^{3} b_{1s} a_{s2} = 52$$

$$b_{23} = \sum_{s=1}^{3} b_{1s} a_{s3} = 54$$

Similarly

$$b_{31} = -4 \qquad b_{32} = 4 \qquad b_{33} = -1{,}458$$

Having these values, the determinant corresponding to Eq. (2.41) can be written

$$D_n(\lambda) = \begin{vmatrix} 2 - \lambda & 4 & -6 \\ 56 - \lambda^2 & 52 & 54 \\ -4 - \lambda^3 & 4 & -1{,}458 \end{vmatrix}$$

The cofactors of elements (1, 1), (2, 1), and (3, 1) can be calculated as follows

$$C_{11} = (52)(-1{,}458) - (54)(4) = -76{,}032$$
$$C_{21} = -((4)(-1{,}458) - (-6)(4)) = 5{,}808$$
$$C_{31} = (4)(54) - (-6)(52) = 528$$

The summation term of Eq. (2.46) can be calculated from the above values

$$-\sum_{i=1}^{n} b_{i1} C_{i1} = -(b_{11} C_{11} + b_{21} C_{21} + b_{31} C_{31})$$

$$= -171{,}072$$

Using the calculated values in Eq. (2.46), we get

$$528\lambda^3 + 5,808\lambda^2 - 76,032\lambda - 171,072 = 0$$

After dividing by the coefficient of λ^3 term, we have

$$\lambda^3 + 11\lambda^2 - 144\lambda - 324 = 0$$

As mentioned before, the roots of this equation can be obtained by root-finding methods of Chap. 3. The characteristic vectors corresponding to these roots are obtained by solving the set of simultaneous equations (2.42).

DETERMINATION OF THE EIGENVECTORS OF A

As we have seen, the eigenvector equation $AX = \lambda X$ can be written in the form

$$[A - \lambda I](X) = K(\lambda)(X) = 0 \tag{2.48}$$

where $K(\lambda)$ is the characteristic matrix of A.

To each eigenvalue $\lambda = \lambda_i$, there corresponds a column vector X_i called an eigenvector. The eigenvector X_i is said to belong to the eigenvalue λ_i; it satisfies the equation

$$K(\lambda_i) X_i = 0 \quad i = 1, 2, 3, \ldots, n \tag{2.49}$$

Equation (2.49) gives nonzero solutions for the numbers in the column X_i since $\det K(\lambda_i) = 0$.

As a consequence of the definition of inverse matrix, we have

$$K(\lambda) \text{ adj } K(\lambda) = \det K(\lambda) I = p(\lambda) I \tag{2.50}$$

If we place $\lambda = \lambda_i$ in Eq. (2.50), we have

$$K(\lambda_i) \text{ adj } K(\lambda_i) = p(\lambda_i) I = 0 \tag{2.51}$$

This result follows from the fact that $p(\lambda_i) = 0$, since λ_i is an eigenvalue of A. If we now compare Eqs. (2.49) and (2.51), we conclude that

X_i can be taken to be proportional to any row of $\operatorname{adj} K(\lambda_i)$; we can then write

$$X_i = (\text{const})(\text{any column of adj } K(\lambda_i)) \tag{2.52}$$

By the use of Eq. (2.52) we again obtain an eigendirection, as was the case in the simple situation of the previously given second-order square matrix.

In that simple case, we had $A = \begin{bmatrix} a & b \\ c & d \end{bmatrix}$; in this case we have

$$K(\lambda) = \begin{bmatrix} a-\lambda & b \\ c & d-\lambda \end{bmatrix} \tag{2.53}$$

We now take the transpose of $K(\lambda)$ and obtain

$$K'(\lambda) = \begin{bmatrix} a-\lambda & c \\ b & d-\lambda \end{bmatrix} \tag{2.54}$$

The adjoint of $K(\lambda)$ is therefore

$$\operatorname{adj} K(\lambda) = \begin{bmatrix} d-\lambda & -b \\ -c & a-\lambda \end{bmatrix} \tag{2.55}$$

Then by the use of Eq. (2.52) we have

$$X_1 = (\text{const}) \begin{bmatrix} d-\lambda \\ -c \end{bmatrix} \quad \text{or} \quad X_1 = (\text{const}) \begin{bmatrix} -b \\ a-\lambda \end{bmatrix} \tag{2.56}$$

Similarly, the eigenvector X_2 is proportional to either one of the columns of $\operatorname{adj} K(\lambda_2)$. This is the same result that we obtained previously.

The result (2.52) is true in general and enables one to obtain the eigenvectors of a matrix whose eigenvalues are distinct. It sometimes happens that the characteristic equation (2.22) has repeated roots. In this case the matrix A is said to have multiple eigenvalues. We shall reserve the discussion of multiple eigenvalues for a later section. For the moment, we will consider the case for which the characteristic equation (2.22) has distinct roots and therefore the matrix A has n distinct eigenvalues $\lambda_1, \lambda_2, \lambda_3, \ldots, \lambda_n$. In such case the eigenvectors or eigendirections are given by Eq. (2.52).

PROPERTIES OF THE EIGENVECTORS OF A SQUARE MATRIX

In this section we shall discuss several of the most important properties of the eigenvectors and eigenvalues of a square matrix A. These properties are of great theoretical and practical importance.

Property 1. Let A be a symmetrical matrix, so that $a_{ij} = a_{ji}$ or $A' = A$. If the eigenvectors X_i and X_j belong to the eigenvalues λ_i and λ_j, where $\lambda_i \neq \lambda_j$, then X_i and X_j are orthogonal vectors. That is, their scalar product vanishes, or

$$X_i'X_j = X_j'X_i = 0 \tag{2.57}$$

By definition, X_i satisfies the eigenvector equation

$$AX_i = \lambda_i X_i \tag{2.58}$$

If we take the transpose of Eq. (2.58) and use the reversal law of transposed products, we obtain

$$X_i'A = \lambda_i X_i' \tag{2.59}$$

If we postmultiply Eq. (2.59) by X_j, we obtain

$$X_i'AX_j = \lambda_i X_i'X_j \tag{2.60}$$

The vector X_j satisfies the eigenvector equation

$$AX_j = \lambda_j X_j \tag{2.61}$$

If we now premultiply Eq. (2.61) by X_i' we obtain

$$X_i' A X_j = \lambda_j X_i' X_j \tag{2.62}$$

If we now subtract Eq. (2.62) from Eq. (2.60), we have

$$0 = (\lambda_i - \lambda_j) X_i' X_j \tag{2.63}$$

Since $\lambda_i \neq \lambda_j$, Eq. (2.63) gives the important relation

$$X_i' X_j = 0 \tag{2.64}$$

Since the scalar product vanishes, X_i and X_j are orthogonal.

Property 2. If A is a real symmetrical matrix (a matrix all of whose elements are real numbers), all of its eigenvalues are real. This is not an obvious fact since the roots of an nth-order equation with real coefficients are, in general, complex numbers.

Let X be a typical eigenvector of A; therefore, we have

$$AX = \lambda X \tag{2.65}$$

The complex conjugate of this equation is

$$AX^* = \lambda^* X^* \tag{2.66}$$

where the symbol * denotes the complex conjugate. The matrix A, being real, is unchanged. The transpose of Eq. (2.66) is

$$X^{*\prime} A = \lambda^* X^{*\prime} \tag{2.67}$$

We now postmultiply Eq. (2.67) by X to obtain

$$X^{*\prime} A X = \lambda^* X^{*\prime} X \tag{2.68}$$

But if we premultiply Eq. (2.65) by $X^{*\prime}$ we have

$$X^{*\prime} A X = \lambda X^{*\prime} X \tag{2.69}$$

If we now subtract Eq. (2.69) from Eq. (2.68) we have

$$0 = (\lambda^* - \lambda) X^{*\prime} X \qquad (2.70)$$

However, if X consists of the elements $x_1, x_2, x_3, \ldots, x_n$, then $X^{*\prime} X$ is the sum of the squares of the absolute values of the elements of X. Hence $X^{*\prime} X$ is a nonzero positive quantity, and therefore to satisfy Eq. (2.70) we must have

$$\lambda^* = \lambda \qquad (2.71)$$

This equation implies that λ is real. Since λ is a typical eigenvalue, all the eigenvalues of A are real numbers.

Property 3. If A is a real, skew-symmetrical matrix, so that $A' = -A$, all its eigenvalues are imaginary. This may be established by the same argument that was used to establish property 2.

Property 4. The eigenvalues of a real orthogonal matrix are all of unit modulus (have absolute value of unity). An orthogonal matrix has the property that $A'A = I$, that is, the inverse of an orthogonal matrix is its transpose. Matrices of this type are of great importance in dynamics and other branches of applied mathematics. Let A be an orthogonal matrix; its eigenvector equation is

$$AX = \lambda X \qquad (2.72)$$

Although A has real elements, λ and X are not necessarily real. The conjugate of this equation is

$$AX^* = \lambda^* X^* \qquad (2.73)$$

where λ^* denotes the conjugate of λ. The transpose of Eq. (2.73) is

$$X^{*\prime} A' = \lambda^* X^{*\prime} \qquad (2.74)$$

If we now postmultiply Eq. (2.74) by $AX = \lambda X$, we obtain

$$X^{*\prime} A' A X = \lambda^* \lambda (X^{*\prime} X) \qquad (2.75)$$

Since A is an orthogonal matrix $A'A = I$, Eq. (2.75) becomes

$$X^{*'}X = \lambda^* \lambda (X^{*'}X) \tag{2.76}$$

If we now divide by the scalar $X^{*'}X$, we obtain

$$\lambda^* \lambda = 1 \tag{2.77}$$

It is therefore evident that the modulus (absolute value) of λ is unity. Since λ is a typical eigenvalue of the orthogonal matrix A, all its eigenvalues are of unit modulus.

The Transform of a Square Matrix (Similarity Transformations). In many applications of matrix algebra and matrix calculus to problems in physics and engineering, it is important to consider the matrix transformation

$$C = B^{-1}AB \tag{2.78}$$

where B is a nonsingular matrix. Such a transformation is called a *similarity* transformation. The matrix C is said to be the transform of A by the matrix B. It is interesting to note that if A is a symmetric matrix, $(A' = A)$ and B is an orthogonal matrix $(B^{-1} = B')$, then the transform of A, C is

$$C = B'AB \tag{2.79}$$

In this case we have by the reversal law of transposed products:

$$C' = (B'AB)' = B'AB = C \tag{2.80}$$

Therefore, since $C' = C$, C is a symmetrical matrix.

Property 5. The eigenvalues of $C = B^{-1}AB$ are equal to the eigenvalues of A. That is, the eigenvalues of the transform matrix C are equal to the eigenvalues of A. To prove this proposition, we note that the characteristic matrix of C is given by

$$(C - \lambda I) = (B^{-1}AB - \lambda I) = B^{-1}(A - \lambda I)B \tag{2.81}$$

Therefore the characteristic polynomial of C, $Q(\lambda)$ is given by

$$Q(\lambda) = \det(C - \lambda I) = \det B^{-1}(\det(A - \lambda I))\det B \qquad (2.82)$$

since $(\det B^{-1}) = 1/\det(B)$, it is apparent from Eq. (2.82) that

$$Q(\lambda) = \det(A - \lambda I) = p(\lambda) \qquad (2.83)$$

where $p(\lambda)$ is the characteristic polynomial of A. Thus, it is evident that the matrices A and C have the same characteristic equations

$$Q(\lambda) = p(\lambda) = 0 \qquad (2.84)$$

Therefore, the eigenvalues of C are equal to the eigenvalues of A.

Property 6. If C is the transform of A so that $C = B^{-1}AB$, the determinant of C is equal to the determinant of A and the trace of C is equal to the trace of A.

To prove this, it is only necessary to recall that the eigenvalues of A and C are the same. Since the determinant of a square matrix is equal to the product of its eigenvalues and the trace of a square matrix is equal to the sum of its eigenvalues, it is evident that A and C have the same determinant and the same trace.

Property 7. The reciprocals of the eigenvalues of A are the eigenvalues of A^{-1} and the eigenvectors of A are the eigenvectors of A^{-1}. To prove this statement, consider the eigenvector equation of A

$$AX = \lambda X \qquad (2.85)$$

Let us premultiply Eq. (2.85) by A^{-1} to obtain

$$A^{-1}AX = IX = \lambda A^{-1}X \qquad (2.86)$$

This equation can be written in the form

$$A^{-1}X = \lambda^{-1}X \qquad (2.87)$$

Equation (2.87) is the eigenvector equation of A^{-1}; therefore, λ^{-1} is the eigenvalue of A^{-1} and X is the eigenvector of A^{-1}. Since λ and X are typical eigenvalues and eigenvectors, it is easy to see that the reciprocals of the eigenvalues of A are the eigenvalues of A^{-1}, and that the eigenvectors of A are the eigenvectors of A^{-1}.

THE CASE OF REPEATED EIGENVALUES

In certain practical cases it happens that the characteristic equation (2.22) has multiple roots; in such a case the matrix A is said to have multiple or repeated eigenvalues.   The question of determining the eigenvectors that belong to a repeated eigenvalue is a difficult one. We shall consider here only the simplest aspects of this question. In order to make matters definite, we shall consider only a third-order case and then make the proper generalizations to cases of higher order.

Let A be a $(3, 3)$ matrix with a triple eigenvalue λ_0. Let X_1, X_2, and X_3 be the eigenvectors that belong to the eigenvalue λ_0. The eigenvector equation for the eigenvalue λ_0 is written as

$$
\begin{aligned}
AX_1 &= \lambda_0 X_1 \\
AX_2 &= \lambda_0 X_2 \\
AX_3 &= \lambda_0 X_3
\end{aligned}
\tag{2.88}
$$

We can multiply the first of the equations (2.88) by a, the second one by b, and the third one by c, add the results, and thus obtain

$$
A[aX_1 + bX_2 + cX_3] = \lambda_0[aX_1 + bX_2 + cX_3]
\tag{2.89}
$$

If we now let

$$
Y = (aX_1 + bX_2 + cX_3)
\tag{2.90}
$$

it is apparent that Eq. (2.89) can be written in the form

$$
AY = \lambda_0 Y
\tag{2.91}
$$

Therefore, Y is an eigenvector of A belonging to the eigenvalue λ_0; hence a linear combination of the eigenvectors belonging to the multiple eigenvalue λ_0 is also an eigenvector belonging to λ_0. This fact can be utilized in obtaining eigenvectors belonging to multiple eigenvalues.

Example. As an example of a symmetrical matrix that has a double eigenvalue, consider the matrix

$$A = \begin{bmatrix} 0 & 1 & 1 \\ 1 & 0 & -1 \\ 1 & -1 & 0 \end{bmatrix} \tag{2.92}$$

This matrix has the following eigenvalues:

$$\lambda_1 = -2 \quad \lambda_2 = 1 \quad \lambda_3 = 1 \tag{2.93}$$

An eigenvector belonging to $\lambda_1 = -2$ can be taken to be

$$X_1 = \begin{bmatrix} -1 \\ 1 \\ 1 \end{bmatrix} \tag{2.94}$$

The eigenvector equation $AX_2 = \lambda_2 X_2$ when written out is

$$\begin{bmatrix} 0 & 1 & 1 \\ 1 & 0 & -1 \\ 1 & -1 & 0 \end{bmatrix} \begin{bmatrix} x_1 \\ x_2 \\ x_3 \end{bmatrix} = \begin{bmatrix} x_1 \\ x_2 \\ x_3 \end{bmatrix} \tag{2.95}$$

This equation if expanded yields three equations of the type

$$x_3 = (x_1 - x_2) \tag{2.96}$$

We can therefore take

$$X_2 = \begin{bmatrix} 1 \\ 1 \\ 0 \end{bmatrix} \tag{2.97}$$

for the eigenvector belonging to the eigenvalue $\lambda_2 = 1$.

In order to obtain the eigenvector X_3 belonging to $\lambda_3 = 1$ we note that the matrix A is a symmetrical matrix and that as a consequence of property 1 above, the eigenvectors of A are mutually orthogonal. Let us make use of Eq. (2.96) in the form

$$x_1 = x_2 + x_3 \tag{2.98}$$

and take $x_2 = h$ and $x_3 = k$ so that the third eigenvector has the form

$$X_3 = \begin{bmatrix} h + k \\ h \\ k \end{bmatrix} \tag{2.99}$$

We now adjust X_3 so that

$$X_1' X_3 = -(h + k) + h + k = 0 \tag{2.100}$$

and

$$X_2' X_3 = (h + k) + h = 0 \quad \text{or} \quad k = -2h \quad (h + k) = -h \tag{2.101}$$

If we substitute these values of k into Eq. (2.99), we obtain

$$X_3 = \begin{bmatrix} -h \\ h \\ -2h \end{bmatrix} = h \begin{bmatrix} -1 \\ 1 \\ -2 \end{bmatrix} \quad h = \text{an arbitrary constant} \tag{2.102}$$

If we let $h = -1$, we finally obtain

$$X_1 = \begin{bmatrix} -1 \\ 1 \\ 1 \end{bmatrix} \quad X_2 = \begin{bmatrix} 1 \\ 1 \\ 0 \end{bmatrix} \quad X_3 = \begin{bmatrix} 1 \\ -1 \\ 2 \end{bmatrix} \tag{2.103}$$

as a set of three orthogonal eigenvectors of the matrix A.

Example. Let it be required to find the eigenvalues and the eigenvectors of the matrix

$$
A = \begin{bmatrix} 7 & 4 & -4 \\ 4 & -8 & -1 \\ -4 & -1 & -8 \end{bmatrix}
\qquad (2.104)
$$

The eigenvalues of A are found to be

$$
\lambda_1 = 9 \qquad \lambda_2 = -9 \qquad \lambda_3 = -9
\qquad (2.105)
$$

The eigenvector X_1 can be found by the standard procedure of Eq. (2.52) to be

$$
X_1 = \begin{bmatrix} 4 \\ 1 \\ -1 \end{bmatrix}
\qquad (2.106)
$$

When $\lambda_2 = \lambda_3 = -9$, the eigenvector equation will be found to give three equations all proportional to the equation

$$
4x_1 + x_2 - x_3 = 0 \quad \text{or} \quad 4x_1 = x_3 - x_2
\qquad (2.107)
$$

If we choose $x_2 = 1$, $x_3 = 1$, then $x_1 = 0$; therefore, we can take X_2 to be

$$
X_2 = \begin{bmatrix} 0 \\ 1 \\ 1 \end{bmatrix}
\qquad (2.108)
$$

If we now take $x_2 = h$, $x_3 = k$, then $x_1 = (k - h)/4$, and we can therefore take

$$X_3 = \begin{bmatrix} \dfrac{k-h}{4} \\ h \\ k \end{bmatrix} \qquad (2.109)$$

Since A is symmetric, X_1, X_2, and X_3 are orthogonal, therefore $X_1' X_3 = 0$, $X_2' X_3 = 0$. We find that $k = -h$ ensures orthogonality, and if we choose $k = 2$, we have

$$X_3 = \begin{bmatrix} 1 \\ -2 \\ 2 \end{bmatrix} \qquad (2.110)$$

Therefore the three eigenvectors of A can be taken to be

$$X_1 = \begin{bmatrix} 4 \\ 1 \\ -1 \end{bmatrix} \quad X_2 = \begin{bmatrix} 0 \\ 1 \\ 1 \end{bmatrix} \quad X_3 = \begin{bmatrix} 1 \\ -2 \\ 2 \end{bmatrix} \qquad (2.111)$$

MATRIX ITERATION METHOD OF EVALUATING EIGENVALUES AND EIGENVECTORS

As discussed above, the characteristic values and vectors of matrix $[A]$ are related as follows:

$$[A - \lambda I](S) = 0 \qquad (2.112)$$

This matrix equation can be written

$$\begin{bmatrix} a_{11} - \lambda & a_{12} & a_{13} \cdots a_{1n} \\ a_{21} & a_{22} - \lambda & a_{23} \cdots a_{2n} \\ \cdots\cdots\cdots\cdots\cdots\cdots\cdots\cdots\cdots\cdots \\ a_{n1} & a_{n2} & a_{n3} \cdots a_{nn} - \lambda \end{bmatrix} \begin{bmatrix} s_1 \\ s_2 \\ \cdot \\ \cdot \\ s_n \end{bmatrix} = \begin{bmatrix} 0 \\ 0 \\ \cdot \\ \cdot \\ 0 \end{bmatrix} \qquad (2.113)$$

where λ is the characteristic value, a's are the elements of matrix $[A]$, and s's are the elements of the characteristic vector corresponding to λ. It is further noted that in many engineering applications the matrix A is a real-valued symmetrical matrix, that is, $a_{ij} = a_{ji}$ and a_{ij}'s are real. From matrix theory discussed above it is known that the characteristic values of real symmetrical matrices are real. This property will be utilized in obtaining characteristic values and vectors by the following iteration process.

Equation (2.113) can also be written in the form

$$\begin{bmatrix} a_{11} & a_{12} & a_{13} \cdots a_{1n} \\ a_{21} & a_{22} & a_{23} \cdots a_{2n} \\ \cdots\cdots\cdots\cdots\cdots\cdots\cdots \\ a_{n1} & a_{n2} & a_{n3} \cdots a_{nn} \end{bmatrix} \begin{bmatrix} s_1 \\ s_2 \\ \cdot \\ \cdot \\ s_n \end{bmatrix} = \lambda \begin{bmatrix} s_1 \\ s_2 \\ \cdot \\ \cdot \\ s_n \end{bmatrix} \qquad (2.114)$$

In the matrix iteration method of finding the value of λ and the corresponding values of $s_1, s_2, \ldots, s_n$ we start with an arbitrary set of $s_1, s_2, \ldots, s_n$, and for convenience in calculations choose $s_n = 1$. (It should be noted that Eq. (2.113) forms a set of homogeneous equations, i.e., equations involving s's with the right side equal to zero. The absolute values of s's cannot be determined. However, ratios of s's can be obtained. By setting $s_n = 1$, in effect we are calculating the ratios of s's to s_n or $s_1/s_n, s_2/s_n, s_3/s_n, \ldots, s_n/s_n$.) Setting these values of s's on the left of Eq. (2.113) and multiplying by the matrix $[A]$, we obtain a column matrix. Dividing this column matrix by the element in the last row s_n, we obtain the next approximation to the value of $\lambda(= s_n)$ and the characteristic vector consisting of $s_1, s_2, \ldots, s_n$. Continuing this process the iteration will converge, resulting in the characteristic value λ and the corresponding characteristic vector [4]. It can be shown that the characteristic value obtained by this method is the largest characteristic value or equivalently the largest root of the characteristic equation.

It can be shown that the rate of convergence of this iteration process depends on the numerical separation of the eigenvalues of matrix A.

Example. As a numerical example of evaluating characteristic values and vectors by the matrix iteration method, consider the following matrix:

$$\begin{bmatrix} 2 & -1 & 0 \\ -1 & 2 & -1 \\ 0 & -1 & 1 \end{bmatrix} \tag{2.115}$$

As a first approximation to s's or the characteristic vector, consider the column matrix

$$\begin{bmatrix} 1 \\ -1 \\ 1 \end{bmatrix} \tag{2.116}$$

Note that the last element is set to unity. It is also generally known that the elements of the characteristic vector corresponding to the highest characteristic value are not all of the same sign. For this reason the sign of the center element is chosen as minus. The iteration method will, of course, converge even if $[1, 1, 1]$ were chosen. However, in this case more iterations will be required. Multiplying the two matrices and keeping the last element of the resulting column matrix as unity, we get

$$\begin{bmatrix} 2 & -1 & 0 \\ -1 & 2 & -1 \\ 0 & -1 & 1 \end{bmatrix} \begin{bmatrix} 1 \\ -1 \\ 1 \end{bmatrix} = 2 \begin{bmatrix} 1.5 \\ -2 \\ 1 \end{bmatrix} \tag{2.117}$$

Using the column matrix on the right as the next approximation and following the above procedure, we get

$$
\begin{bmatrix} 2 & -1 & 0 \\ -1 & 2 & -1 \\ 6 & -1 & 1 \end{bmatrix} \begin{bmatrix} 1.5 \\ -2 \\ 1 \end{bmatrix} = 3 \begin{bmatrix} 1.6666 \\ -2.1666 \\ 1 \end{bmatrix}
$$

$$
\begin{bmatrix} \text{same as above} \end{bmatrix} \begin{bmatrix} 1.6666 \\ -2.1666 \\ 1 \end{bmatrix} = 3.1666 \begin{bmatrix} 1.7368 \\ -2.2105 \\ 1 \end{bmatrix} \qquad (2.118)
$$

$$
\begin{bmatrix} \text{same as above} \end{bmatrix} \begin{bmatrix} 1.7368 \\ -2.2105 \\ 1 \end{bmatrix} = 3.2105 \begin{bmatrix} 1.7701 \\ -2.2295 \\ 1 \end{bmatrix}
$$

Continuing this process, iterations 10 and 11 will result in

$$
3.247 \begin{bmatrix} 1.802 \\ -2.247 \\ 1.0 \end{bmatrix} \quad \text{and} \quad 3.247 \begin{bmatrix} 1.802 \\ -2.247 \\ 1.0 \end{bmatrix} \qquad (2.119)
$$

It is seen that iterations converge and the values at iterations 10 or 11 represent the largest characteristic value (3.247) and the corresponding vector 1.802, −2.247, 1.0. It is sometimes desirable to normalize the characteristic vector to a value of unity. If s_1, s_2, and s_3 are the components of the characteristic vector, this results:

$$
s_1{}^2 + s_2{}^2 + s_3{}^2 = 1 \qquad (2.120)
$$

Or in the case of the above example, all the components should be divided by the square root of

$$
(1.802)^2 + (2.247)^2 + (1.0)^2 = 9.2962 \qquad (2.121)
$$

or 3.049. The normalized characteristic vector then becomes

$$
\begin{bmatrix} \dfrac{1.802}{3.049} \\[2ex] -\dfrac{2.247}{3.049} \\[2ex] \dfrac{1.0}{3.049} \end{bmatrix} = \begin{bmatrix} 0.591 \\[2ex] -0.736 \\[2ex] 0.328 \end{bmatrix}
\qquad\qquad (2.122)
$$

FORTRAN PROGRAM FOR EVALUATION OF CHARACTERISTIC VALUES AND VECTORS

The following FORTRAN program [1] has been written for the iteration process discussed above. The data entered in the program are the same as the above example. The output of the program consists of 20 iterations and is given following the listing of the program.

Statement 10 reserves computer storage for indexed variables. Statements 20 and 30 read the order of the matrix ($N = 3$) and the values of its elements $S(I, J)$. Statement 40 reads the initial characteristic vector $P(J)$ as $(1, -1, 1)$. K in statement 50 is the number of the iteration and is indexed by one at each iteration. Statements 70 through 110 multiply the matrix by the initial characteristic vector. Statements 120 through 140 change the designation of the characteristic vector from $Z(I)$ to $P(I)$ to be used in future iterations. Statements 150 through 180 print out the iteration number and the characteristic value which is equal to $P(N)$. Statements 190 through 210 divide the characteristic vector by $P(N)$, as was done in the example problem, to keep the last element of this vector equal to unity. Statements 220 and 230 print out the characteristic vector. In statement 240 the number of iterations is arbitrarily set at 20. If K is less than or equal to 20, the loop starting with statement 60 is repeated and another iteration is obtained. The computer results are shown following the program. It is seen that the characteristic vector and the corresponding characteristic value to three significant figures remain unchanged after iteration number 10.

```
C        PROGRAM TO EVALUATE CHARACTERISTIC VALUE AND VECTOR
   10    DIMENSION S(100,100),P(100),Z(100)
   20    DATA N/3/
   30    DATA ((S(I,J),J=1,3),I=1,3)/2.,−1.,0.,−1.,2.,−1.,0.,−1.,1./
   40    DATA (P(J),J=1,3)/1.,−1.,1./
   50    K=0
   60    K=K+1
   70    DO 110 I=1,N
   80    Z (I)=0.
   90    DO 110J=1,N
  100    Z(I)=S(I,J)*P(J)+Z(I)
  110    CONTINUE
  120    DO 140 I=1,N
  130    P(I)=Z(I)
  140    CONTINUE
  150    WRITE (2,160) K
  160    FORMAT (18H ITERATION NUMBER= I5)
  170    WRITE (2,180) P(N)
  180    FORMAT (22H CHARACTERISTIC VALUE= 1PF10.3)
  190    DO 210 I=1, N
  200    P(I)=P(I)/P(N)
  210    CONTINUE
  220    WRITE (2,230) (P(I), I=1,3)
  230    FORMAT (22H CHARACTERISTIC VECTOR /(6X1PE10.3))
  240    IF (K-20) 60,250,250
  250    STOP
  260    END
```

Computer solution obtained:

```
ITERATION NUMBER =              1
CHARACTERISTIC VALUE =          2.000E 00
CHARACTERISTIC VECTOR
         1.500E 00
        −2.000E 00
         1.000E 00
ITERATION NUMBER =              2
CHARACTERISTIC VALUE =          3.000E 00
CHARACTERISTIC VECTOR
         1.667E 00
        −2.211E 00
         1.000E 00
ITERATION NUMBER =              3
CHARACTERISTIC VALUE =          3.167E 00
CHARACTERISTIC VECTOR
```

```
        1.737E 00
       −2.211E 00
        1.000E 00
ITERATION NUMBER =            4
CHARACTERISTIC VALUE =            3.211E 00
CHARACTERISTIC VECTOR
        1.770E 00
       −2.230E 00
        1.000E 00
ITERATION NUMBER =            5
CHARACTERISTIC VALUE =            3.230E 00
CHARACTERISTIC VECTOR
        1.787E 00
       −2.239E 00
        1.000E 00
ITERATION NUMBER =            6
CHARACTERISTIC VALUE =            3.239E 00
CHARACTERISTIC VECTOR
        1.795E 00
       −2.243E 00
        1.000E 00
ITERATION NUMBER =            7
CHARACTERISTIC VALUE =            3.243E 00
CHARACTERISTIC VECTOR
        1.798E 00
       −2.245E 00
        1.000E 00
ITERATION NUMBER =            8
CHARACTERISTIC VALUE =            3.245E 00
CHARACTERISTIC VECTOR
        1.800E 00
       −2.246E 00
        1.000E 00
ITERATION NUMBER =            9
CHARACTERISTIC VALUE =            3.246E 00
CHARACTERISTIC VECTOR
        1.801E 00
       −2.247E 00
        1.000E 00
ITERATION NUMBER =            10
CHARACTERISTIC VALUE =            3.247E 00
CHARACTERISTIC VECTOR
        1.802E 00
       −2.247E 00
        1.000E 00
```

ITERATION NUMBER =   11
CHARACTERISTIC VALUE =   3.247E 00
CHARACTERISTIC VECTOR
        1.802E 00
       −2.247E 00
        1.000E 00
ITERATION NUMBER =   12
CHARACTERISTIC VALUE =   3.247E 00
CHARACTERISTIC VECTOR
        1.802E 00
       −2.247E 00
        1.000E 00
ITERATION NUMBER =   13
CHARACTERISTIC VALUE =   3.247E 00
CHARACTERISTIC VECTOR
        1.802E 00
       −2.247E 00
        1.000E 00
ITERATION NUMBER =   14
CHARACTERISTIC VALUE =   3.247E 00
CHARACTERISTIC VECTOR
        1.802E 00
       −2.247E 00
        1.000E 00
ITERATION NUMBER =   15
CHARACTERISTIC VALUE =   3.247E 00
CHARACTERISTIC VECTOR
        1.802E 00
       −2.247E 00
        1.000E 00
ITERATION NUMBER =   16
CHARACTERISTIC VALUE =   3.247E 00
CHARACTERISTIC VECTOR
        1.802E 00
       −2.247E 00
        1.000E 00
ITERATION NUMBER =   17
CHARACTERISTIC VALUE =   3.247E 00
CHARACTERISTIC VECTOR
        1.802E 00
       −2.247E 00
        1.000E 00
ITERATION NUMBER =   18
CHARACTERISTIC VALUE =   3.247E 00
CHARACTERISTIC VECTOR

```
        1.802E 00
       -2.247E 00
        1.000E 00
ITERATION NUMBER =          19
CHARACTERISTIC VALUE =          3.247E 00
CHARACTERISTIC VECTOR
        1.802E 00
       -2.247E 00
        1.000E 00
ITERATION NUMBER =          20
CHARACTERISTIC VALUE =          3.247E 00
CHARACTERISTIC VECTOR
        1.802E 00
       -2.247E 00
        1.000E 00
```

ITERATION METHOD FOR OBTAINING THE SMALLEST CHARACTERISTIC VALUES AND VECTORS

The method described above gives the highest characteristic value and the corresponding characteristic vector. However, in many applications the lowest characteristic value and the corresponding vector are desired. To obtain this, we divide Eq. (2.114) by λ and premultiply both sides by matrix $[a_{ij}]^{-1}$. Denoting $1/\lambda = \omega$, we get

$$\omega \begin{bmatrix} s_1 \\ s_2 \\ \vdots \\ s_n \end{bmatrix} = \begin{bmatrix} a_{11} & a_{12} & a_{13} \cdots a_{1n} \\ a_{21} & a_{22} & a_{23} \cdots a_{2n} \\ \cdots\cdots\cdots\cdots\cdots \\ a_{n1} & a_{n2} & a_{n3} \cdots a_{nn} \end{bmatrix}^{-1} \begin{bmatrix} s_1 \\ s_2 \\ \vdots \\ s_n \end{bmatrix} \qquad (2.123)$$

We note that the iteration process for obtaining ω and $s_1, s_2, s_3, \ldots, s_n$ remains the same as before with matrix $[a_{ij}]$ replaced by its inverse $[a_{ij}]^{-1}$ as given by Eq. (2.123). From this equation the value of ω will correspond to the largest characteristic value of the inverse matrix $[a_{ij}]^{-1}$ or the smallest characteristic value of $[a_{ij}]$ since $\omega = 1/\lambda$.

Example. The inverse of the matrix of the previous example can be obtained, for example by augmented matrix method, as written below

$$\begin{bmatrix} 2 & -1 & 0 \\ -1 & 2 & -1 \\ 0 & -1 & 1 \end{bmatrix}^{-1} = \begin{bmatrix} 1 & 1 & 1 \\ 1 & 2 & 2 \\ 1 & 2 & 3 \end{bmatrix}$$

Using this inverse together with an initial iteration characteristic vector

$$\begin{bmatrix} 1 \\ 1 \\ 1 \end{bmatrix}$$

and proceeding with the iteration processs as was done in the above example, we get

First iteration

$$\begin{bmatrix} 1 & 1 & 1 \\ 1 & 2 & 2 \\ 1 & 2 & 3 \end{bmatrix} \begin{bmatrix} 1 \\ 1 \\ 1 \end{bmatrix} = 6 \begin{bmatrix} 0.50 \\ 0.8333 \\ 1.0 \end{bmatrix}$$
....   ⋮   ⋮

Fifth iteration

$$\begin{bmatrix} 1 & 1 & 1 \\ 1 & 2 & 2 \\ 1 & 2 & 3 \end{bmatrix} \begin{bmatrix} 0.4451 \\ 0.8020 \\ 1.0 \end{bmatrix} = 5.049 \begin{bmatrix} 0.4451 \\ 0.8020 \\ 1.0 \end{bmatrix}$$
....   ⋮   ⋮

Tenth iteration

$$\begin{bmatrix} 1 & 1 & 1 \\ 1 & 2 & 2 \\ 1 & 2 & 3 \end{bmatrix} \begin{bmatrix} 0.4450 \\ 0.8019 \\ 1.0 \end{bmatrix} = 5.049 \begin{bmatrix} 0.4450 \\ 0.8019 \\ 1.0 \end{bmatrix}$$

From this, the smallest characteristic value will be $\lambda = 1/5.049 = 0.198$ with the corresponding components of the characteristic vector 0.445, 0.8019, and 1.0.

This vector can be put in normalized form by dividing the components by the square root of the sum of the squares. This results in the values 0.328, 0.591, and 0.736 (normalized) with the lowest characteristic value of $\lambda = 0.198$.

USE OF ORTHOGONALITY RELATIONS IN OBTAINING CHARACTERISTIC VALUES AND VECTORS

After obtaining the largest characteristic value and the corresponding characteristic vector it is possible to use the orthogonality relations to obtain other characteristic values and vectors. (This method is also used when the characteristic equation has double roots, i.e., two characteristic values are identical.) From matrix theory we have that the characteristic vectors of real symmetric matrices are orthogonal. That is, if $s_1, s_2, \ldots, s_n$ are the components of the characteristic vector corresponding to the characteristic value λ, and $s_1', s_2', s_3', \ldots, s_n'$ are the components of the characteristic vector corresponding to λ', then

$$s_1 s_1' + s_2 s_2' + \cdots + s_n s_n' = 0 \tag{2.124}$$

Assuming that we have obtained the largest characteristic value λ, and the corresponding vector $s_1, s_2, \ldots, s_n$, we can obtain the next characteristic value and vector λ' and $s_1', s_2', \ldots, s_n'$ by writing Eqs. (2.114) in the form

$$\begin{bmatrix} a_{11} & a_{12} & \cdots & a_{1n} \\ a_{21} & a_{22} & \cdots & a_{2n} \\ \cdots\cdots\cdots\cdots\cdots \\ a_{n1} & a_{n2} & \cdots & a_{nn} \end{bmatrix} \begin{bmatrix} s_1' \\ s_2' \\ \vdots \\ s_n' \end{bmatrix} = \lambda' \begin{bmatrix} s_1' \\ s_2' \\ \vdots \\ s_n' \end{bmatrix} \tag{2.125}$$

Solving Eq. (2.124) for s_n', we get

$$s_n' = -\frac{s_1}{s_n} s_1' - \frac{s_2}{s_n} s_2' - \frac{s_3}{s_n} s_3' - \cdots - \frac{s_{n-1}}{s_n} s_{n-1}' \tag{2.126}$$

Writing matrix equation (2.125) in the form of linear equations and substituting the value of s'_n from Eq. (2.126), we get a set of n equations in $(n - 1)$ unknowns $s'_1, s'_2, \ldots, s'_{n-1}$.

Since the last equation is redundant (it will always be satisfied) we can drop this equation and write

$$\begin{bmatrix} b_{11} & b_{12} & \cdots & b_{1,\,n-1} \\ b_{21} & b_{22} & \cdots & b_{2,\,n-1} \\ \cdots\cdots\cdots\cdots\cdots\cdots\cdots\cdots\cdots \\ b_{n-1,\,1} & b_{n-1,\,2} & \cdots & b_{n-1,\,n-1} \end{bmatrix} \begin{bmatrix} s'_1 \\ s'_2 \\ \vdots \\ s'_{n-1} \end{bmatrix} = \lambda' \begin{bmatrix} s'_1 \\ s'_2 \\ \vdots \\ s'_{n-1} \end{bmatrix} \qquad (2.127)$$

where b's are the new coefficients. These equations can be solved by the iteration process, discussed previously, for λ' and $s'_1, s'_2, \ldots, s'_{n-1}$. Having these values, Eq. (2.126) will give the value of s'_n.

Additional characteristic values and vectors can be obtained by writing the orthogonality relations between the vectors already calculated and vectors whose values are desired, and using these relations to reduce the number of equations.

Example. Consider the previously given example problem where the largest characteristic value and the corresponding vector of the matrix

$$\begin{bmatrix} 2 & -1 & 0 \\ -1 & 2 & -1 \\ 0 & -1 & 1 \end{bmatrix}$$

were found by iteration method. These values are 3.247 and 1.802, -2.247, 1.0. It is desired, using the orthogonality condition, to find the next largest characteristic value λ' and the corresponding vector with components s'_1, s'_2 and s'_3.

The orthogonality condition of Eq. (2.126) results in

$$s'_3 = -\frac{s_1}{s_3} s'_1 - \frac{s_2}{s_3} s'_2 \qquad (2.128)$$

$$= -1.802 s'_1 + 2.247 s'_2$$

The values of s_1', s_2', and s_3' should also satisfy Eq. (2.125):

$$\begin{bmatrix} 2 & -1 & 0 \\ -1 & 2 & -1 \\ 0 & -1 & 1 \end{bmatrix} \begin{bmatrix} s_1' \\ s_2' \\ s_3' \end{bmatrix} = \lambda' \begin{bmatrix} s_1' \\ s_2' \\ s_3' \end{bmatrix} \qquad (2.129)$$

Writing the above relation in equation form and substituting for s_3' in terms of s_1' and s_2' in the first two of these equations, we get

$$(2 - \lambda')s_1' - s_2' = 0$$
$$0.802s_1' + (-0.247 - \lambda')s_2' = 0 \qquad (2.130)$$
$$-s_2' + (1 - \lambda')s_3' = 0$$

The last equation of (2.130) is not needed as the first two equations are sufficient to solve for s_1', s_2', and λ'. Putting these in matrix form, we get

$$\begin{bmatrix} 2 & -1 \\ 0.802 & -0.247 \end{bmatrix} \begin{bmatrix} s_1' \\ s_2' \end{bmatrix} = \lambda' \begin{bmatrix} s_1' \\ s_2' \end{bmatrix} \qquad (2.131)$$

The above relations, which correspond to Eq. (2.127), can be solved by matrix iteration,* discussed previously, by assuming that $s_1' = 1.0$ and $s_2' = 1.0$ for the first iteration. The iteration process will converge to three significant figures at the sixth iteration, resulting in

$$\lambda' = 1.555$$
$$s_1' = 2.247 \qquad (2.132)$$
$$s_2' = 1.0$$

Using the above values in the expression for s_3' in terms of s_1' and s_2', we get

---

*Since in this case only two equations are involved, they can also be solved simultaneously.

$$s_3' = -1.802s_1' + 2.247s_2'$$
$$= -1.802 \times 2.247 + 2.247 \times 1.0 \qquad (2.133)$$
$$= -1.802$$

The value of s_3' can be set equal to 1.0, as was done in the case of highest characteristic value and vector, by dividing the components of the new characteristic vector by -1.802. This results in

$$s_1' = \frac{2.247}{-1.802} = -1.247$$

$$s_2' = \frac{1.0}{-1.802} = -0.555 \qquad (2.134)$$

$$s_3' = \frac{-1.802}{-1.802} = 1.0$$

with $\lambda' = 1.555$. The above values can be normalized by dividing each component by the square root of the sum of the squares. This results in

$$s_1' = -0.736 \qquad s_2' = -0.328 \qquad s_3' = 0.591$$

In order to calculate the next characteristic value and the corresponding vector we can proceed as follows. Assume the desired values are λ'' and s_1'', s_2'', and s_3''. Write the orthogonality relations between these and the vectors S and S' as given in Eq. (2.124).

$$1.802s_1'' - 2.247s_2'' + s_3'' = 0$$
$$-1.247s_1'' - 0.555s_2'' + s_3'' = 0 \qquad (2.135)$$

Solving these equations for s_3'' and s_2'' in terms of s_1'', we get

$$s_2'' = 1.802s_1''$$
$$s_3'' = 2.247s_1'' \qquad (2.136)$$

Substituting the above relations for the values of s_2'' and s_3'' in the first of Eqs. (2.129) written for s'',

$$
\begin{bmatrix} 2 & -1 & 0 \\ -1 & 2 & -1 \\ 0 & -1 & 1 \end{bmatrix} \begin{bmatrix} s_1'' \\ s_2'' \\ s_3'' \end{bmatrix} = \lambda'' \begin{bmatrix} s_1'' \\ s_2'' \\ s_3'' \end{bmatrix} \tag{2.137}
$$

we get

$$ 2s_1'' - s_2'' + 0s_3'' = \lambda'' s_1'' \tag{2.138} $$

or

$$ 2s_1'' - 1.802s_1'' = \lambda'' s_1'' \tag{2.139} $$

Setting the value of s_1'' in the above equation equal to unity, we get

$$ \lambda'' = 0.198 \tag{2.140} $$

And, from the previously given equations,

$$
\begin{aligned}
s_2'' &= 1.802 \\
s_3'' &= 2.247
\end{aligned}
\tag{2.141}
$$

The above characteristic vector can be written in the form where $s_3'' = 1.0$ by dividing the components of the vector by $s_3'' = 2.247$. This results in

$$ s_1'' = 0.445 \qquad s_2'' = 0.802 \qquad s_3'' = 1.0 $$

These values compare with previously obtained results for the lowest characteristic value. The normalized form of this characteristic vector can be obtained by dividing the above values by the square root of the sum of the squares. This gives

$$ s_1'' = 0.328 \qquad s_2'' = 0.591 \qquad s_3'' = 0.736 \tag{2.142} $$

An additional method for the computation of the characteristic equations and the corresponding vectors will be discussed in the following section. This method, referred to as the method of A. M. Danilevsky, is based primarily on the properties of similarity transformation discussed previously.

THE METHOD OF A. M. DANILEVSKY FOR OBTAINING THE CHARACTERISTIC EQUATION

The method of Danilevsky for obtaining the characteristic equation is based on the reduction of matrix A, whose characteristic equation is desired,

$$A = \begin{bmatrix} a_{11} & a_{12} & a_{13} \cdots a_{1n} \\ a_{21} & a_{22} & a_{23} \cdots a_{2n} \\ \cdots\cdots\cdots\cdots\cdots\cdots\cdots\cdots \\ a_{n1} & a_{n2} & a_{n3} \cdots a_{nn} \end{bmatrix} \qquad (2.143)$$

to matrix P of the form

$$P = \begin{bmatrix} p_1 & p_2 \cdots p_{n-1} & p_n \\ 1 & 0 \cdots 0 & 0 \\ 0 & 1 \cdots 0 & 0 \\ \cdots\cdots\cdots\cdots\cdots\cdots \\ 0 & 0 \cdots 1 & 0 \end{bmatrix} \qquad (2.144)$$

where the elements $p_1, p_2, p_3, \ldots, p_n$ represent the coefficients of the characteristic equation of matrix A.

It can be shown that the transformation of matrix A to matrix P can be accomplished by a series of similarity transformations. Thus, on the strength of the theory that the characteristic equations of similar matrices are identical, the characteristic equations of matrices A and P will be the same. The fact that the elements of matrix P represent the coefficients of the characteristic polynomial is readily seen by writing the characteristic determinant of P as

$$
P(\lambda) =
\begin{vmatrix}
p_1 - \lambda & p_2 & p_3 & \cdots & p_n \\
1 & -\lambda & 0 & \cdots & 0 \\
0 & 1 & -\lambda & \cdots & 0 \\
\multicolumn{5}{c}{\dotfill} \\
0 & 0 & 0 & \cdots & -\lambda
\end{vmatrix}
\tag{2.145}
$$

and expanding it in terms of the elements of the first row

$$
D(\lambda) = (-1)^n \left(\lambda^n - p_1 \lambda^{n-1} - \cdots - p_n \right) = 0
\tag{2.146}
$$

where $D(\lambda)$ is the desired characteristic polynomial and n is the order of the square matrix A.

We demonstrate the transformation from matrix A to matrix P for a 4×4 matrix

$$
A =
\begin{vmatrix}
a_{11} & a_{12} & a_{13} & a_{14} \\
a_{21} & a_{22} & a_{23} & a_{24} \\
a_{31} & a_{32} & a_{33} & a_{34} \\
a_{41} & a_{42} & a_{43} & a_{44}
\end{vmatrix}
\tag{2.147}
$$

Postmultiplying Eq. (2.147) by the matrix M, which is formed from the elements of the fourth row of the matrix A, shown below

$$
M =
\begin{bmatrix}
1 & 0 & 0 & 0 \\
0 & 1 & 0 & 0 \\
-\dfrac{a_{41}}{a_{43}} & -\dfrac{a_{42}}{a_{43}} & \dfrac{1}{a_{43}} & -\dfrac{a_{44}}{a_{43}} \\
0 & 0 & 0 & 1
\end{bmatrix}
\tag{2.148}
$$

will result in

$$AM = \begin{bmatrix} b_{11} & b_{12} & b_{13} & b_{14} \\ b_{21} & b_{22} & b_{23} & b_{24} \\ b_{31} & b_{32} & b_{33} & b_{34} \\ 0 & 0 & 1 & 0 \end{bmatrix}$$

(2.149)

where the elements b of the above are

$$b_{11} = a_{11} - \frac{a_{13}a_{41}}{a_{43}}, \quad b_{12} = a_{12} - \frac{a_{13}a_{42}}{a_{43}}, \quad b_{13} = \frac{a_{13}}{a_{43}}, \dots$$

$$b_{21} = a_{21} - \frac{a_{23}a_{41}}{a_{43}}, \quad b_{22} = a_{22} - \frac{a_{23}a_{42}}{a_{43}}, \dots$$

(2.150)

Note that the last row of the matrix of Eq. (2.149) is the same as that of the P matrix of Eq. (2.144). The matrix AM of Eq. (2.149) is not as yet similar to matrix A of Eq. (2.147). It can be made similar to A by premultiplying by the inverse of matrix M. Thus, the similarity transformation of A into the form of Eq. (2.149) is written as

$$C = M^{-1}AM$$

(2.151)

where M^{-1} is readily obtained from Eq. (2.148) as

$$M^{-1} = \begin{bmatrix} 1 & 0 & 0 & 0 \\ 0 & 1 & 0 & 0 \\ a_{41} & a_{42} & a_{43} & a_{44} \\ 0 & 0 & 0 & 1 \end{bmatrix}$$

(2.152)

The third row of this matrix is the same as the last row of the matrix of Eq. (2.147). The premultiplication of matrix AM of Eq. (2.149) by M^{-1}, Eq. (2.151), does not affect the last row of the matrix of Eq. (2.149).

At this point the matrix C of Eq. (2.151) will be of the form

$$C = \begin{bmatrix} c_{11} & c_{12} & c_{13} & c_{14} \\ c_{21} & c_{22} & c_{23} & c_{24} \\ c_{31} & c_{32} & c_{33} & c_{34} \\ 0 & 0 & 1 & 0 \end{bmatrix} \tag{2.153}$$

The process can be continued in reducing the above matrix to the form

$$D = \begin{bmatrix} 1 & 0 & 0 & 0 \\ c_{31} & c_{32} & c_{33} & c_{34} \\ 0 & 0 & 1 & 0 \\ 0 & 0 & 0 & 1 \end{bmatrix} \begin{bmatrix} c_{11} & c_{12} & c_{13} & c_{14} \\ c_{21} & c_{22} & c_{23} & c_{24} \\ c_{31} & c_{32} & c_{33} & c_{34} \\ 0 & 0 & 1 & 0 \end{bmatrix} \tag{2.154}$$

$$\begin{bmatrix} 1 & 0 & 0 & 0 \\ -\dfrac{c_{31}}{c_{32}} & \dfrac{1}{c_{32}} & -\dfrac{c_{33}}{c_{32}} & -\dfrac{c_{44}}{c_{32}} \\ 0 & 0 & 1 & 0 \\ 0 & 0 & 0 & 1 \end{bmatrix} = \begin{bmatrix} d_{11} & d_{12} & d_{13} & d_{14} \\ d_{21} & d_{22} & d_{23} & d_{24} \\ 0 & 1 & 0 & 0 \\ 0 & 0 & 1 & 0 \end{bmatrix}$$

Carrying the above process a step further will reduce the 4×4 matrix of Eq. (2.147) to the form of Eq. (2.144), that is,

$$P = \begin{bmatrix} p_1 & p_2 & p_3 & p_4 \\ 1 & 0 & 0 & 0 \\ 0 & 1 & 0 & 0 \\ 0 & 0 & 1 & 0 \end{bmatrix} \tag{2.155}$$

where p's represents the coefficients of the characteristic polynomial of Eq. (2.146).

As an example of the above procedure, consider the 4×4 matrix

$$A = \begin{bmatrix} -5.50988 & 1.87009 & 0.422908 & 0.008814 \\ 0.287865 & -11.8117 & 5.7119 & 0.058717 \\ 0.049099 & 4.30803 & -12.9707 & 0.229326 \\ 0.006235 & 0.269851 & 1.39737 & -17.5962 \end{bmatrix} \qquad (2.156)$$

whose characteristic equation is required. Following the above procedure the matrix M of the similarity transformation is calculated as

$$M = \begin{bmatrix} 1 & 0 & 0 & 0 \\ 0 & 1 & 0 & 0 \\ -0.00462 & -0.193113 & 0.71563 & 12.5924 \\ 0 & 0 & 0 & 1 \end{bmatrix} \qquad (2.157)$$

The matrix M^{-1} is calculated as given by Eq. (2.152). That is,

$$M^{-1} = \begin{bmatrix} 1 & 0 & 0 & 0 \\ 0 & 1 & 0 & 0 \\ 0.006235 & 0.269851 & 1.39737 & -17.5962 \\ 0 & 0 & 0 & 1 \end{bmatrix} \qquad (2.158)$$

Having matrices M and M^{-1}, we implement the similarity transformation of Eq. (2.151)

$$C = M^{-1}AM = \begin{bmatrix} -5.51177 & 1.78842 & 0.302646 & 5.33423 \\ 0.262379 & -12.9147 & 4.08761 & 71.9851 \\ 0.185919 & 6.04616 & -29.462 & -208.456 \\ 0 & 0 & 1 & 0 \end{bmatrix}$$

$$(2.159)$$

At this point we set matrix C above identical to matrix A and repeat the similarity transformation about the third row of the above matrix. The M matrix for this case becomes

$$
M = \begin{bmatrix}
1 & 0 & 0 & 0 \\
-0.03075 & 0.165394 & 4.87284 & 34.4774 \\
0 & 0 & 1 & 0 \\
0 & 0 & 0 & 1
\end{bmatrix}
\qquad (2.160)
$$

The inverse of the M matrix can be obtained from the third row of the C matrix as

$$
M^{-1} = \begin{bmatrix}
1 & 0 & 0 & 0 \\
0.185919 & 6.04616 & -29.462 & -208.456 \\
0 & 0 & 1 & 0 \\
0 & 0 & 0 & 1
\end{bmatrix}
\qquad (2.161)
$$

The resulting matrix after the similarity transformation becomes

$$
M^{-1}CM = \begin{bmatrix}
-5.56676 & 0.295794 & 9.01733 & 66.9944 \\
2.95252 & -42.3217 & -562.559 & -2244.47 \\
0 & 1 & 0 & 0 \\
0 & 0 & 1 & 0
\end{bmatrix}
\qquad (2.162)
$$

We can set the above matrix identical to A and repeat the procedure. This, after obtaining matrices M, M^{-1}, and $M^{-1}CM$, will result in the matrix of the coefficients

$$
P = \begin{bmatrix}
-47.8885 & -797.281 & -5349.48 & -12296.6 \\
1 & 0 & 0 & 0 \\
0 & 1 & 0 & 0 \\
0 & 0 & 1 & 0
\end{bmatrix}
\qquad (2.163)
$$

From the above matrix and Eq. (2.146), the characteristic equation of the problem can be written

$$D(\lambda) = (-1)^4(\lambda^4 + 47.8885\lambda^3 + 797.281\lambda^2$$
$$+ 5349.48\lambda + 12296.6) = 0$$

(2.164)

The roots of this equation can be obtained by the methods described in Chap. 3.

COMPUTER PROGRAM FOR THE METHOD OF A. M. DANILEVSKY

The following computer program has been written in BASIC programming language (see [8]). This programming language includes matrix operations denoted by MAT at the start of each statement. The program starts with dimension statement 100 reserving space for matrices A, M, I (inverse of M), and $P = AM$. Statement 110 specified the order of the matrix whose characteristic equation is required. Note that in the BASIC language the first row and first columns of a matrix are denoted by row zero and column zero, respectively. For this reason $N = 3$ specifies a fourth-order matrix. Statement 120 reads the elements of matrix A, from left to right, as given in data statements 350 through 380. Statements 160 through 180 set matrices M, I, and P equal to identity matrix, zero matrix, and zero matrix, respectively. Statements 190 through 220 form the similarity transformation matrix M. Statement 250 obtains $I = M^{-1}$ matrix. Statement 280 obtains the $P = AM$ matrix while statement 290 calculates $M^{-1} AM$ matrix and denotes this as the A matrix for the second iteration which starts from statement number 160 and so on.

The results given for the example problem were obtained through the above described computer program.

In exceptional cases the process of similarity transformation, as described above, may result in a matrix where a zero may appear in the position of an element whose value should be transformed to unity in the next iteration. Consider the $C = M^{-1} AM$ matrix at the end of some iteration to have the form

```
100   DIM A (10,10),M(10,10),I(10,10),P(10,10)
110   LET N=3
120   MAT READ A(N,N)
130   PRINT "MATRIX A"
140   MAT PRINT A
150   LET L=N-1
160   MAT M=IDN (N,N)
170   MAT I=ZER (N,N)
180   MAT P=ZER (N,N)
190   FOR J=0 TO N
200   LET M(L,J)= -A(L+1,J)/A(L+1,L)
210   NEXT J
220   LET M(L,L)=1/A(L+1,L)
230   PRINT "MATRIX M"
240   MAT PRINT M
250   MAT I=INV (M)
260   PRINT "INVERSE OF MATRIX M"
270   MAT PRINT I
280   MAT P=A*M
290   MAT A=I*P
300   PRINT "--------- MATRIX A ----------"
310   MAT PRINT A
320   LET L=L -1
330   IF L < 0 THEN 390
340   GO TO 160
350   DATA -5.50988, 1.87009, .422908, .008814
360   DATA .287865, -11.8117, 5.7119, .058717
370   DATA .049099, 4.30803, -12.9707, .229326
380   DATA .006235, .269851, 1.39737, -17.5962
390   END
```

$$C = \begin{bmatrix} c_{11} & c_{12} \cdots & c_{1k} & & \cdots & c_{1n} \\ c_{21} & c_{22} \cdots & c_{2k} & & \cdots & c_{2n} \\ \multicolumn{6}{c}{\dotfill} \\ c_{k1} & c_{k2} \cdots & c_{kk} & & \cdots & c_{kn} \\ 0 & 0 \ldots & 1 & 0 & \ldots & 0 \\ 0 & 0 \ldots & 0 & 1 & \ldots & 0 \\ \multicolumn{6}{c}{\dotfill 1} \end{bmatrix} \qquad (2.165)$$

with the element $c_{k, k-1} = 0$. In this case a nonzero element is chosen along the same row, c_{ki} $(i < k - 1)$ and the ith and $k - 1$ columns and simultaneously the ith and $k - 1$ rows are interchanged. This operation will be equivalent to a similarity transformation with the M matrix

$$
M = \begin{bmatrix} 1 \\ & \ddots \\ & & 1 \\ & & 0 \text{-----} 1 & & & i\text{th row} \\ & & | 1 & & | \\ & & | & \ddots & | \\ & & | & & 1 | \\ & & 1 \text{-----} 0 & & & k - 1 \text{ row} \\ & & & | 1 \\ & i\text{th} & & | & \ddots \\ & \text{col.} & & | \\ & & & k - 1 & 1 & & 1 \\ & & & \text{col.} \end{bmatrix} \tag{2.166}
$$

The above matrix has the property $M^2 = I$, or $M = M^{-1}$; therefore, postmultiplication and premultiplication of matrix C of Eq. (2.165) by M and M^{-1}, respectively, will represent a similarity transformation with no change to the characteristic equation.

CHARACTERISTIC VECTOR BY THE METHOD OF A. M. DANILEVSKY

Let λ be one of the characteristic values of matrix A obtained from the characteristic polynomial derived above. The characteristic vector corresponding to λ is denoted by $X = (x_1, x_2, x_3, \ldots, x_n)$ where small x's are the components of the vector. The characteristic vector corresponding to the matrix P is denoted by $Y = (y_1, y_2, \ldots, y_n)$. It can be shown that

$$
X = M_1 M_2 \cdots M_{n-1} Y \tag{2.167}
$$

where $M_1, M_2, \ldots, M_{n-1}$ are, respectively, the transformations which reduce the matrix A of Eq. (2.143) to matrix P of Eq. (2.144). The components of the characteristics vector of Eq. (2.144) can readily be obtained from

$$\begin{bmatrix} p_1 - \lambda & p_2 & p_3 & \cdots & p_n \\ 1 & -\lambda & 0 & \cdots & 0 \\ 0 & 1 & -\lambda & \cdots & 0 \\ \multicolumn{5}{c}{\cdots\cdots\cdots\cdots\cdots\cdots\cdots\cdots} \\ 0 & 0 & 0 & \cdots & -\lambda \end{bmatrix} \begin{bmatrix} y_1 \\ y_2 \\ y_3 \\ \vdots \\ y_n \end{bmatrix} = 0$$

which in equation form becomes

$$(p_1 - \lambda_1)y_1 + p_2 y_2 + \cdots + p_n y_n = 0$$
$$y_1 - \lambda y_2 \qquad\qquad = 0$$
$$\cdots\cdots\cdots\cdots\cdots\cdots\cdots\cdots$$
$$y_{n-1} - \lambda y_n \qquad\qquad = 0$$

Setting $y_n = 1$ we get $y_{n-1} = \lambda$, $y_{n-2} = \lambda^2$, $y_1 = \lambda^{n-1}$. Having these values and the product of M transformations, Eq. (2.167) will result in the characteristic vector corresponding to λ.

At this point a comparison of the previously discussed matrix iteration method and the method of A. M. Danilevsky is in order. In the method of Danilevsky, the characteristic equation is obtained and then the roots of this equation, the characteristic values, are calculated by numerical root finding methods of Chap. 3. Thus, the numerical proximity of the roots does not increase the number of required calculations as was the case in matrix iteration method. On the other hand, the matrix iteration method produces the largest eigenvalue and the corresponding eigenvector at the same time. Since in many engineering problems the highest (or the lowest, which can be obtained by the inversion of the matrix of the coefficients) eigenvalue and the corresponding eigenvector of systems represented by real symmetrical matrices are required, the matrix iteration method under these conditions is preferred.

The number of multiplications and divisions by the method of A. M. Danilevsky in obtaining the characteristic equation is $(n - 1)(n^2 + n - 1)$ where n is the order of the matrix. The method of A. N. Krylov for obtaining the characteristic polynomial requires a greater number of arithmetical operations than Danilevsky's method. It is used primarily because of the simplicity of computational procedure. Another shortcoming of Krylov's method comes from the fact that arithmetical operations are to be performed on numbers of different orders of magnitude, resulting, in many cases, in loss of accuracy. In cases where the characteristic equation has multiple roots, Krylov's method results in the coefficients of lower-degree polynomial.

The iteration methods of calculating eigenvectors and eigenvalues discussed above were primarily given from a computational point of view. Characteristic values and vectors, of course, have definite physical meanings in vibration problems as described in the following pages.

CHARACTERISTIC VALUES AND VECTORS IN MECHANICAL VIBRATIONS

It will be shown in the following example that the characteristic values correspond to the natural frequencies of vibration in mechanical systems while the characteristic vectors correspond to the normal modes of vibration. If a given conservative (no friction) dynamical system is initially displaced proportional to one of its normal modes, the system will vibrate indefinitely with a frequency equal to the natural frequency of the system corresponding to that normal mode.

Consider the dynamical system of Fig. 2.1. The moment of inertia of each disk is denoted by I while the torsional stiffness of the connecting shaft is denoted by c. The length of the shaft between disks is l. The equations of motion of each disk from Newton's second law of motion can be written

$$-I\ddot{q}_1 = \frac{c}{l} q_1 + \frac{c}{l}(q_1 - q_2)$$

$$-I\ddot{q}_2 = \frac{c}{l}(q_1 - q_2) + \frac{c}{l}(q_2 - q_3) \qquad (2.168)$$

$$-I\ddot{q}_3 = \frac{c}{l}(q_3 - q_2)$$

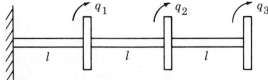

figure 2-1 *Vibrating shaft with three disks.*

where q's are the angular displacements and dots represent derivatives with respect to time. The above equations can also be written in the form

$$-I\ddot{q}_1 = \frac{c}{l}(2q_1 - q_2)$$

$$-I\ddot{q}_2 = \frac{c}{l}(-q_1 + 2q_2 - q_3) \qquad (2.169)$$

$$-I\ddot{q}_3 = \frac{c}{l}(-q_2 + q_3)$$

Consider the solution of the above equations in the form

$$q_i = A_i \sin(\omega t + \psi) \qquad i = 1, 2, 3 \qquad (2.170)$$

Obtaining the second derivative of the above equation and substituting in Eq. (2.169), we obtain

$$2A_1 - A_2 \qquad = \left(\frac{Il}{c}\omega^2\right)A_1$$

$$-A_1 + 2A_2 - A_3 = \left(\frac{Il}{c}\omega^2\right)A_2 \qquad (2.171)$$

$$- A_2 + A_3 = \left(\frac{Il}{c}\omega^2\right)A_3$$

The above homogeneous equation can have nonzero solutions for A_1, A_2, and A_3 only if the following determinant vanishes

$$\begin{vmatrix} 2 - \lambda & -1 & 0 \\ -1 & 2 - \lambda & -1 \\ 0 & -1 & 1 - \lambda \end{vmatrix} = 0 \qquad (2.172)$$

where

$$\lambda = \frac{Il}{c}\omega^2 \qquad (2.173)$$

With this definition of λ Eqs. (2.171) can be written in matrix notation as follows

$$\begin{bmatrix} 2 & -1 & 0 \\ -1 & 2 & -1 \\ 0 & -1 & 1 \end{bmatrix} \begin{bmatrix} A_1 \\ A_2 \\ A_3 \end{bmatrix} = \lambda \begin{bmatrix} A_1 \\ A_2 \\ A_3 \end{bmatrix} \tag{2.174}$$

The roots of the determinant of Eq. (2.172) were previously calculated to be

$$\lambda_1 = 3.238 \quad \lambda_2 = 1.555 \quad \lambda_3 = 0.198 \tag{2.175}$$

Using these values in Eq. (2.174) will result in corresponding values of A's, which were also previously calculated as

$$\lambda_1 = 3.238 \begin{cases} A_1 = 0.591 \\ A_2 = -0.736 \\ A_3 = 0.328 \end{cases}$$

$$\lambda_2 = 1.555 \begin{cases} A_1 = -0.736 \\ A_2 = -0.328 \\ A_3 = 0.591 \end{cases} \tag{2.176}$$

$$\lambda_3 = 0.198 \begin{cases} A_1 = 0.328 \\ A_2 = 0.591 \\ A_3 = 0.736 \end{cases}$$

Note that the angular frequency ω can be calculated from the values of λ's using Eq. (2.173); thus

$$\omega_1 = 1.801 \sqrt{\frac{c}{Il}}$$

$$\omega_2 = 1.247 \sqrt{\frac{c}{Il}} \qquad (2.177)$$

$$\omega_3 = 0.445 \sqrt{\frac{c}{Il}}$$

Using the above values and the values A from Eq. (2.176) in Eq. (2.170), we obtain for the complete solution

$$
\begin{aligned}
q_1 = {} & 0.591A \sin (\omega_1 t + \psi_1) - 0.736B \sin (\omega_2 t + \psi_2) \\
& + 0.328C \sin (\omega_3 t + \psi_3) \\
q_2 = {} & -0.736A \sin (\omega_1 t + \psi_1) - 0.328B \sin (\omega_2 t + \psi_2) \qquad (2.178) \\
& + 0.591C \sin (\omega_3 t + \psi_3) \\
q_3 = {} & 0.328A \sin (\omega_1 t + \psi_1) + 0.591B \sin (\omega_2 t + \psi_2) \\
& + 0.736C \sin (\omega_3 t + \psi_3)
\end{aligned}
$$

It should be remembered that the A's of Eq. (2.176) represent the ratio of characteristic values rather than their absolute values. For this reason the constants A, B, and C are added to the above equations.

In dynamics the values of ω as given by Eq. (2.177) are referred to as the natural frequencies of vibration and corresponding values of A's of Eq. (2.176) are modes of vibration.

It can be shown that an initial displacement of the disks of Fig. 2-1 proportional to the values of the modes of vibration (Eq. (2.176)) will result in oscillations of the system at the rate given by the corresponding frequency.

To prove the above statement, assume the following initial conditions for the system of equations (2.178):

$$
\begin{array}{llll}
\text{at } t = 0 & q_1 = 0.591 & q_2 = -0.736 & q_3 = 0.328 \\
& \dot{q}_1 = 0 & \dot{q}_2 = 0 & \dot{q}_3 = 0
\end{array} \qquad (2.179)
$$

Using the derivative condition above in Eq. (2.178) we obtain the following set of equations

$$
\begin{bmatrix}
0.591 & -0.736 & 0.328 \\
-0.736 & -0.328 & 0.591 \\
0.328 & 0.591 & 0.736
\end{bmatrix}
\begin{bmatrix}
A\omega_1 \cos \psi_1 \\
B\omega_2 \cos \psi_2 \\
C\omega_3 \cos \psi_3
\end{bmatrix} = 0
\qquad (2.180)
$$

Since the determinant of the coefficients of the above equation is not zero, the only way that the equation will be satisfied is

$$
\cos \psi_1 = 0 \quad \psi_1 = 90° \quad \text{and} \quad \psi_2 = 90° \quad \psi_3 = 90° \qquad (2.181)
$$

Using the initial condition for q_1, q_2, and q_3 from Eqs. (2.179) and (2.178), we get

$$
\begin{bmatrix}
0.591 & -0.736 & 0.328 \\
-0.736 & -0.328 & 0.591 \\
0.328 & 0.591 & 0.736
\end{bmatrix}
\begin{bmatrix}
A \\
B \\
C
\end{bmatrix} =
\begin{bmatrix}
0.591 \\
-0.736 \\
0.328
\end{bmatrix}
\qquad (2.182)
$$

From the above equation

$$
A = 1 \quad B = 0 \quad C = 0 \qquad (2.183)
$$

And, the system of equations (2.178) becomes

$$
\begin{aligned}
q_1 &= 0.591 \cos \omega_1 t \\
q_2 &= -0.736 \cos \omega_1 t \\
q_3 &= 0.328 \cos \omega_1 t
\end{aligned}
\qquad (2.184)
$$

Thus, the system will vibrate with the frequency ω_1 once displaced by the normal modes corresponding to this frequency.

HE RELATION BETWEEN THE EIGENVECTORS
ND EIGENVALUES OF A MATRIX AND THOSE
F ITS TRANSPOSE

In this section we will consider the relation of the eigenvalues and igenvectors of a matrix A and those of its transpose A'.

Let the eigenvector equation of the matrix A be

$$AX = \lambda X \tag{2.185}$$

et the eigenvector equation of A' be

$$A'Y = \mu Y \tag{2.186}$$

The characteristic equation of the matrix A is

$$P(\lambda) = \det (A - \lambda I) = 0 \tag{2.187}$$

The characteristic equation of A' is

$$Q(\lambda) = \det (A' - \mu I) = 0 \tag{2.188}$$

However, since interchanging the rows and the columns of a determiant does not affect its value, we have

$$P(\lambda) = \det (A - \lambda I) = \det (A - \lambda I)' = \det (A' - \lambda I) \tag{2.189}$$

ve see that the form of the characteristic equation of A is the same as he characteristic equation of A' and therefore that A and A' have the ame eigenvalues.

If we premultiply Eq. (2.185) by Y', we obtain

$$Y'AX = \lambda Y'X \tag{2.190}$$

f we now take the transpose of Eq. (2.186) and postmultiply the result y X, we obtain

$$Y'AX = \mu Y'X \tag{2.191}$$

We now subtract Eq. (2.191) from Eq. (2.190); the result is

$$(\lambda - \mu) Y'X = 0 \tag{2.192}$$

If the eigenvalues λ and μ are such that $\lambda \neq \mu$, we have the result

$$Y'X = 0 \tag{2.193}$$

It is evident from Eq. (2.193) that the eigenvectors of A are orthogonal to the eigenvectors of A' belonging to different eigenvalues. This fact may be utilized in the determination of the eigenvectors of nonsymmetrical matrices that belong to repeated eigenvalues.

THE GRAM-SCHMIDT ORTHOGONALIZATION PROCESS

In certain applications such as the determination of the eigenvector belonging to repeated eigenvalues, we are concerned with the problem of obtaining an orthogonal set of vectors from any given set of n linearly independent vectors. The procedure by means of which this object is achieved is known in the mathematical literature as the Gram-Schmidt orthogonalization process.

In this process we are given a linearly independent set of vectors $X_1, X_2, X_3, \ldots, X_n$. The problem is to find scalars c_{ij} such that the vectors given by the scheme

$$
\begin{aligned}
Y_1 &= X_1 \\
Y_2 &= c_{21} X_1 + X_2 \\
Y_3 &= c_{31} X_1 + c_{32} X_2 + X_3 \\
&\cdots\cdots\cdots\cdots\cdots\cdots\cdots\cdots\cdots\cdots\cdots\cdots\cdots \\
Y_n &= c_{n1} X_1 + c_{n2} X_2 + \cdots + c_{n,n-1} X_{n-1} + X_n
\end{aligned}
\tag{2.194}
$$

are orthogonal.

Let us introduce the notation

$$(Y_k, X_j) = Y'_k X_j = Y_k \cdot X_j \tag{2.195}$$

so that (Y_k, X_j) denotes the scalar product of the vectors Y_k and X_j. Now consider the vectors

$$Y_1 = X_1$$

$$Y_2 = X_2 - Y_1 \frac{(Y_1, X_2)}{(Y_1, Y_1)}$$

$$Y_3 = X_3 - Y_1 \frac{(Y_1, X_3)}{(Y_1, Y_1)} - Y_2 \frac{(Y_2, X_3)}{(Y_2, Y_2)} \qquad (2.196)$$

$$\dots\dots\dots\dots\dots\dots\dots\dots\dots\dots\dots\dots\dots$$

$$Y_n = X_n - Y_1 \frac{(Y_1, X_n)}{(Y_1, Y_1)} - \dots - Y_{n-1} \frac{(Y_{n-1}, X_n)}{(Y_{n-1}, Y_{n-1})}$$

We note that

$$(Y_1, Y_2) = (X_1, X_2) - \frac{(X_1, X_1)(X_1, X_2)}{(X_1, X_1)} = 0 \qquad (2.197)$$

We can show similarly that $(Y_1, Y_3) = 0$ and that in general

$$(Y_j, Y_k) = 0 \quad \text{for} \quad j \ne k \qquad (2.198)$$

Therefore the vectors Y_1, Y_2, ..., Y_n form an orthogonal set of vectors.
Example. Consider the set of vectors

$$X_1 = (1, 0, 1, 0)$$
$$X_2 = (1, 1, 3, 0)$$
$$X_3 = (0, 2, 0, 1)$$

The orthogonal set of vectors Y are obtained using Eqs. (2.196)

$$Y_1 = (1, 0, 1, 0)$$

$$Y_2 = (1, 1, 3, 0) - \frac{(1, 1, 3, 0), (1, 0, 1, 0)}{(1, 0, 1, 0), (1, 0, 1, 0)} (1, 0, 1, 0)$$

$$= (1, 1, 3, 0) - 2(1, 0, 1, 0)$$

$$= (-1, 1, 1, 0)$$

$$Y_3 = (0, 2, 0, 1) - \frac{(0, 2, 0, 1), (-1, 1, 1, 0)}{(-1, 1, 1, 0), (-1, 1, 1, 0)} (-1, 1, 1, 0)$$

$$- \frac{(0, 2, 0, 0), (1, 0, 1, 0)}{(1, 0, 1, 0), (1, 0, 1, 0)} (1, 0, 1, 0)$$

$$= \left(\frac{2}{3}, \frac{4}{3}, -\frac{2}{3}, 1 \right)$$

THE MODAL MATRIX AND THE SPECTRAL MATRIX

Let the nth-order matrix A have n distinct eigenvalues λ_k, $k = 1, 2, \ldots, n$ and n eigenvectors X_k, $k = 1, 2, 3, \ldots, n$. Let the n eigenvectors X_k be arranged as the columns of a square matrix M. This square matrix is called the modal matrix of A; it has the form

$$M = [X_1 \quad X_2 \quad X_3 \ldots X_n] \tag{2.199}$$

The modal matrix M is therefore a partitioned matrix whose columns are the eigenvectors of the matrix A.

The spectral matrix of the matrix A is a diagonal matrix S of the nth order whose elements are the eigenvalues of the matrix A (we are considering here the case in which the matrix A has n distinct eigenvalues). The spectral matrix S has the form

$$S = \begin{bmatrix} \lambda_1 & 0 & 0 & \cdots & 0 \\ 0 & \lambda_2 & 0 & \cdots & 0 \\ 0 & 0 & \lambda_3 & \cdots & 0 \\ \cdots & \cdots & \cdots & \cdots & \cdots \\ 0 & 0 & 0 \cdots 0 \cdots & \lambda_n \end{bmatrix} = \text{diag} (\lambda_1, \lambda_2, \ldots, \lambda_n) \tag{2.200}$$

The typical eigenvector and eigenvalue X_k and λ_k of the square matrix satisfies the eigenvector equation

$$AX_k = \lambda_k X_k \quad k = 1, 2, 3, \ldots, n \tag{2.201}$$

Let the matrix A be postmultiplied by its modal matrix M to obtain the result

$$AM = A[X_1 \quad X_2 \quad X_3 \ldots X_n] = [AX_1 \quad AX_2 \quad AX_3 \ldots AX_n]$$
$$= [\lambda_1 X_1 \quad \lambda_2 X_2 \quad \lambda_3 X_3 \ldots \lambda_n X_n] = MS \tag{2.202}$$

The matrix M has an inverse provided that the eigenvectors of the matrix A, X_k, $k = 1, 2, 3, \ldots, n$ are linearly independent. This will be the case if A has n distinct eigenvalues λ_k or if A is a symmetrical matrix so that $A' = A$. If the matrix A is not symmetrical and some of its eigenvalues are multiple eigenvalues, the matrix M may not have an inverse. The result (2.202) $AM = MS$ can be premultiplied by M^{-1} in the case M has an inverse. We thus obtain

$$M^{-1}AM = S \tag{2.203}$$

Equation (2.203) is said to represent the diagonalization of the matrix to a *diagonal matrix*. The equation $AM = MS$ can also be postmultiplied by the inverse of the modal matrix M^{-1} to obtain

$$A = MSM^{-1} \tag{2.204}$$

It is interesting to note that if Eq. (2.204) is multiplied by itself, the result is

$$A^2 = (MSM^{-1})(MSM^{-1}) = MS(M^{-1}M)SM^{-1} = MS^2 M^{-1} \tag{2.205}$$

The matrix S^2 is a diagonal matrix whose elements are the squares of the eigenvalues of A. If we multiply Eq. (2.204) by itself k times we obtain

$$A^k = MS^k M^{-1} \quad k = 0, 1, 2, 3, \ldots \tag{2.206}$$

The matrix S^k is a diagonal matrix whose elements are the kth powers of the eigenvalues of A.

THE DIAGONAL FORM OF A SYMMETRICAL
MATRIX WITH DISTINCT EIGENVALUES

Let us consider A to be a symmetrical matrix with distinct eigenvalues. We now multiply the equation $AM = MS$ on the left by the transpose of M, M', to obtain

$$M'AM = M'MS \tag{2.207}$$

We now take the transpose of Eq. (2.207), realizing that $A' = A$, and obtain

$$M'AM = SM'M \tag{2.208}$$

From Eqs. (2.207) and (2.208) it follows that

$$M'MS = SM'M \tag{2.209}$$

Since S is a diagonal matrix, in order to satisfy Eq. (2.209) we must have $M'M =$ a diagonal matrix. If the eigenvectors X_k are normalized to unit length so that

$$X'_r X_s = 1 \quad \text{for} \quad r \neq s \tag{2.210}$$

the modal matrix of the symmetric matrix A has the property that

$$M'M = 1 \tag{2.211}$$

where I is the nth-order unit matrix. Thus M is orthogonal in this case and the Eq. (2.204) takes the form

$$A = MSM' \tag{2.212}$$

In Eq. (2.212) M is the modal matrix of A and M' is its transpose. This equation is the diagonal form of a symmetric matrix with distinct eigenvalues.

Diagonalization of a matrix is closely connected with the problem of obtaining quadratic forms. In many engineering applications, such as finding of the principal axis of inertia and coordinate transformations, in s desirable to reduce a given expression with cross-product terms to a pure quadratic form, that is, to an expression involving squares of the variables without cross-product terms. It will be shown through the following example that this problem is similar to finding the appropriate characteristic values and vectors of the expression with cross-product terms as described above.

Example. Obtain the pure quadratic form for the expression

$$Q = 2x_1^2 - 2x_1 x_2 + 2x_2^2 - 2x_2 x_3 + x_3^2 \tag{2.213}$$

The above relation can be put in symmetrical matrix form as follows:

$$Q = \begin{bmatrix} 2x_1 x_2 & -x_1 x_2 & 0 \\ -x_2 x_1 & 2x_2 x_2 & -x_2 x_3 \\ 0 & -x_3 x_2 & x_3 x_3 \end{bmatrix} \tag{2.214}$$

n the above, it is assumed that $x_i x_j = x_j x_i$. Thus the resulting $[A]$ matrix of the coefficients becomes

$$A = \begin{bmatrix} 2 & -1 & 0 \\ -1 & 2 & -1 \\ 0 & -1 & 1 \end{bmatrix} \tag{2.215}$$

Equation (2.213) or (2.214) can be written in the form

$$Q = (x_1 \quad x_2 \quad x_3) \begin{bmatrix} 2 & -1 & 0 \\ -1 & 2 & -1 \\ 0 & -1 & 1 \end{bmatrix} \begin{bmatrix} x_1 \\ x_2 \\ x_3 \end{bmatrix} \tag{2.216}$$

Multiplying the last two matrices, we get

$$Q = (x_1 \quad x_2 \quad x_3) \begin{bmatrix} 2x_1 - x_2 \\ -x_1 + 2x_2 - x_3 \\ -x_2 + x_3 \end{bmatrix} \tag{2.217}$$

which upon multiplication results in

$$
\begin{aligned}
Q &= x_1(2x_1 - x_2) + x_2(-x_1 + 2x_2 - x_3) + x_3(-x_2 + x_3) \\
&= 2x_1^2 - 2x_1 x_2 + 2x_2^2 - 2x_2 x_3 + x_3^2
\end{aligned} \tag{2.218}
$$

Equation (2.218) is identical to the expression for Q given by Eq. (2.213). This example was purposely chosen to yield matrix [A] whose characteristic values and vectors were previously calculated and are given in the normalized form as follows:

$\lambda_1 = 3.238$;   [0.591, −0.736, 0.328]   first column matrix [M]

$\lambda_2 = 1.555$;   [−0.736, −0.328, 0.591] second column matrix [M]

$\lambda_3 = 0.198$;   [0.328, 0.591, 0.736]   third column matrix [M]

Using these values, the corresponding [M] matrix of Eq. (2.199) becomes

$$M = \begin{bmatrix} 0.591 & -0.736 & 0.328 \\ -0.736 & -0.328 & 0.591 \\ 0.328 & 0.591 & 0.736 \end{bmatrix} \tag{2.219}$$

And, since [M] is an orthogonal matrix $[M' = M^{-1}]$ the corresponding relation to Eq. (2.203) becomes

$$
\begin{bmatrix} 0.591 & -0.736 & 0.328 \\ -0.736 & -0.328 & 0.591 \\ 0.328 & 0.591 & 0.736 \end{bmatrix}
\begin{bmatrix} 2 & -1 & 0 \\ -1 & 2 & -1 \\ 0 & -1 & 1 \end{bmatrix}
$$

$$
\begin{bmatrix} 0.591 & -0.736 & 0.328 \\ -0.736 & -0.328 & 0.591 \\ 0.328 & 0.591 & 0.736 \end{bmatrix}
=
\begin{bmatrix} 3.238 & 0 & 0 \\ 0 & 1.555 & 0 \\ 0 & 0 & 0.198 \end{bmatrix}
\tag{2.220}
$$

Consider the following transformation from x to ξ coordinate system

$$
\begin{bmatrix} x_1 \\ x_2 \\ x_3 \end{bmatrix} = \begin{bmatrix} 0.591 & -0.736 & 0.328 \\ -0.736 & -0.328 & 0.591 \\ 0.328 & 0.591 & 0.736 \end{bmatrix} \begin{bmatrix} \xi_1 \\ \xi_2 \\ \xi_3 \end{bmatrix} \tag{2.221}
$$

In matrix notation this becomes

$$
(x) = [M](\xi) \tag{2.222}
$$

Transposing this, we get

$$
(x)' = [\xi]'(M)' \quad \text{(reversed order)} \tag{2.223}
$$

We now write Eq. (2.223) in full matrix form:

$$
(x_1, x_2, x_3) = (\xi_1, \xi_2, \xi_3) \begin{bmatrix} 0.591 & -0.736 & 0.328 \\ -0.736 & -0.328 & 0.591 \\ 0.328 & 0.591 & 0.736 \end{bmatrix} \tag{2.224}
$$

Using the expressions given for matrices X and X' in Eqs. (2.221) and (2.224) in Eq. (2.216), we get:

$$
Q = (\xi_1, \xi_2, \xi_3) \begin{bmatrix} 0.591 & -0.736 & 0.328 \\ -0.736 & -0.328 & 0.591 \\ 0.328 & 0.591 & 0.736 \end{bmatrix} \begin{bmatrix} 2 & -1 & 0 \\ -1 & 2 & -1 \\ 0 & -1 & 0 \end{bmatrix}
$$
$$
\begin{bmatrix} 0.591 & -0.736 & 0.328 \\ -0.736 & -0.328 & 0.591 \\ 0.328 & 0.591 & 0.736 \end{bmatrix} \begin{bmatrix} \xi_1 \\ \xi_2 \\ \xi_3 \end{bmatrix} \tag{2.225}
$$

Substituting the diagonal matrix of Eq. (2.220) in place of the three matrices at the center of the above expression, we get

$$Q = 3.238\xi_1^2 + 1.555\xi_2^2 + 0.198\xi_3^2$$

which represents the quadratic form of Eq. (2.213) with the transformations of coordinates x to ξ as given by Eq. (2.221). The transformation from the coordinate system ξ to x can easily be obtained by premultiplying Eq. (2.222) by $[M]^{-1} (= [M]')$; this gives

$$\xi = M'x \qquad (2.226)$$

PROBLEMS

1. Find the eigenvalues and eigenvectors of the matrices

$$A = \begin{bmatrix} 0 & 0 & 1 \\ 8 & 0 & 2 \\ 0 & -1 & 5 \end{bmatrix} \quad B = \begin{bmatrix} -2 & -1 & 0 \\ 1 & 2 & 3 \\ 4 & 5 & 6 \end{bmatrix} \quad C = \begin{bmatrix} 1 & 0 & 0 \\ 0 & 1 & 0 \\ 0 & 0 & 1 \end{bmatrix}$$

2. Given the matrix

$$A = \begin{bmatrix} 2 & -2 & 3 \\ 1 & 1 & 1 \\ 1 & 3 & -1 \end{bmatrix}$$

show that its eigenvalues are
$\lambda_1 = 1, \lambda_2 = -2, \lambda_3 = 3$

3. Given the matrix

$$A = \begin{bmatrix} 2 & -2 & 3 \\ 10 & -4 & 5 \\ 5 & -4 & 6 \end{bmatrix}$$

show that $\lambda_1 = 1, \lambda_2 = 1, \lambda_3 = 2$

4. Given the matrix

$$A = \begin{bmatrix} 7 & 4 & -1 \\ 4 & 7 & -1 \\ -4 & -4 & 4 \end{bmatrix}$$

show that $\lambda_1 = 3, \lambda_2 = 3$, and $\lambda_3 = 12$

5. Find the characteristic polynomial of Prob. 2 by the method of A. N. Krylov.

6. Find the characteristic polynomial of Prob. 3 by the method of A. N. Krylov.

7. Find the characteristic polynomial of Prob. 4 by the method of A. N. Krylov.

8. Find the highest characteristic value and the corresponding vector for the matrix

$$\begin{bmatrix} 1 & 1 & 1 \\ 1 & 3 & 2 \\ 1 & 2 & 2 \end{bmatrix}$$

Answer: characteristic value 5.042 vector (5 iterations) 0.555, 1.246, 1,000

by matrix iteration.

9. Find the smallest characteristic value of the above problem by iteration.

10. Using orthogonality relations find all of the characteristic values and vectors of Prob. 2.

11. Obtain the characteristic values and vectors of the matrix

$$\begin{bmatrix} 1 & -4 & 2 \\ -4 & 1 & -2 \\ 2 & -2 & 2 \end{bmatrix}$$

by iteration.

12. Obtain the characteristic equation and characteristic vectors of Prob. 2 by the method of A. M. Danilevsky.

13. Obtain the characteristic equation and characteristic vectors of Prob. 3 by the method of A. M. Danilevsky.

14. Obtain the characteristic polynomial and characteristic vectors of Prob. 4 by the method of A. M. Danilevsky.

15. Obtain the characteristic equation of Prob. 8 by the method of A. M. Danilevsky.

16. Solve the cubic equation obtained above and use the method of Danilevsky to obtain the characteristic vectors.

17. Given the matrix

$$A = \begin{bmatrix} \cos(\theta) & \sin(\theta) \\ -\sin(\theta) & \cos(\theta) \end{bmatrix}$$

show that A is orthogonal, $A'A = I$, and determine the inverse of A and its eigenvalues and eigenvectors.

18. Given the three linearly independent vectors,

$$X_1 = \begin{bmatrix} 4 \\ 5 \\ 0 \end{bmatrix} \quad X_2 = \begin{bmatrix} -1 \\ 0 \\ 5 \end{bmatrix} \quad X_3 = \begin{bmatrix} 3 \\ 6 \\ 8 \end{bmatrix}$$

by the use of the Gram-Schmidt process construct a set of three orthogonal vectors Y_k, $k = 1, 2, 3$ from X_1, X_2, X_3.

19. Reduce the matrix

$$A = \begin{bmatrix} 78 & -60 & 15 \\ 150 & -117 & 30 \\ 200 & -160 & 43 \end{bmatrix}$$

to the diagonal from $M^{-1} AM = S$.

20. Given the matrix

$$A = \begin{bmatrix} 5 & -1 & -1 \\ 1 & 3 & 1 \\ -2 & 2 & 4 \end{bmatrix}$$

show that the eigenvalues are 2, 4, 6. Determine the eigenvectors of A.

21. Show that the eigenvalues of

$$
A = \begin{bmatrix} a & a^2 & 0 \\ \dfrac{r(r+1)}{2} & 0 & a^2 \\ 0 & \dfrac{r(r+1)}{2} & a \end{bmatrix}
$$

are a, $(1 + r)a$, $-ra$. Find the corresponding eigenvectors.

22. Show that the matrix

$$
A = \begin{bmatrix} 0 & 0 & 0 & 1 \\ -1 & 0 & 0 & 0 \\ 0 & -1 & 0 & 0 \\ 0 & 0 & -1 & 0 \end{bmatrix}
$$

is orthogonal. Find its eigenvalues and verify that its eigenvectors are "vectors of zero length."

23. Obtain the quadratic form of the equation

$$
Q = 2x_1{}^2 + 2x_2{}^2 - 15x_3{}^2 + 8x_1 x_2 - 12x_2 x_3 - 12x_1 x_3
$$

and the corresponding transformations.

Hint: The characteristic values of the resulting matrix are -2, -18, and 9.

REFERENCES

1. McCracken, D. D.: "A Guide to FORTRAN IV Programming," John Wiley & Sons, Inc., New York, 1965.

2. Page, Lowell J., and J. Dean Swift: "Elements of Linear Algebra," Chaps. 10 and 11, Ginn and Company, Boston, Mass., 1961.

3. Sokolnikoff, I. S.: "Tensor Analysis," Chap. 1, John Wiley & Sons, Inc., New York, 1951.

4. Von Karman, Theodore, and Maurice A. Biot: "Mathematical Methods in Engineering," Chap. 5, McGraw-Hill Book Company, New York, 1940.

5. Frazer, R. A., W. J. Duncan, and A. R. Collar: "Elementary Matrices," Cambridge University Press, New York (reprinted 1960).

6. Kunz, Kaiser S.: "Numerical Analysis," McGraw-Hill Book Company, New York, 1957.

7. Faddeev, D. K., and V. M. Faddeeva: "Computational Methods of Linear Algebra," W. H. Freeman and Co., Publishers, San Francisco, 1963.

8. BASIC Language, reference manual, General Electric Company, May, 1966.

9. Bellman, Richard: "Introduction to Matrix Analysis," McGraw-Hill Book Company, New York, 1960.

3 ROOTS OF POLYNOMIAL AND ALGEBRAIC EQUATIONS, LAGRANGE'S INTERPOLATION FORMULA, METHOD OF LEAST SQUARES

The formulation of many engineering problems results in linear constant coefficient differential equations, the solutions of which can be obtained by writing the characteristic equation of the differential equation and solving for the roots of the resulting polynomial equation. In addition, the natural frequency of vibration in many electrical and mechanical systems is obtained by solving the characteristic equation of the system, which usually is an algebraic polynomial. In control systems, the stability of the system is determined by location of roots of the denominator of the transfer function of the system. In all of these cases, one is faced with solving for the roots of a polynomial equation. In a number of applications, solutions will result only by finding the roots of algebraic equations which, in addition to powers of the independent variable x, may include trigonometric and/or exponential functions (transcendental equations).

The inverse of finding the roots of polynomial equations, namely, given a number of data points, find a polynomial which passes through those data points, also occurs. In this respect, Lagrange's interpolation formula is the one most often used. However, in problems where the data may include random measurement errors, the method of least squares is often employed.

ROOTS OF POLYNOMIAL EQUATIONS

Consider the nth-order polynomial equation $P_n(x)$ of the independent variable x

$$f(x) = P_n(x) = A_n x^n + A_{n-1} x^{n-1} + \cdots + A_1 x + A_0 = 0 \tag{3.1}$$

where the A's are the coefficients of powers of the independent variable x.

Several numerical methods are available for determining the roots of this polynomial equation. However, before describing these methods a number of algebraic theorems dealing with the roots of polynomials are given for reference. The proofs of these theorems can be obtained in standard texts on algebra.

Theorem 3-1. If $(x - r)$ is a factor of $f(x)$, then r is a root of the equation $f(x) = 0$.

Theorem 3-2. If the polynomial $f(x)$ is divided by $(x - r)$, the remainder will be $f(r)$. For the proof, let $f(x)$ be divided by $(x - r)$ and call the resulting polynomial $Q(x)$ and the remainder R. This gives

$$f(x) = (x - r)Q(x) + R \tag{3.2}$$

for $x = r$

$$f(r) = R$$

Theorem 3-3. If r is a root of the equation $f(x) = 0$, then $(x - r)$ is a factor of the polynomial $f(x)$.

Theorem 3-4. A polynomial of degree n cannot have more than n roots.

Theorem 3-5. If $f(x)$ is a polynomial and a and b are real numbers such that $f(a) > 0$ and $f(b) < 0$, then $f(x)$ has a root between a and b.

Theorem 3-6. The coefficients of the polynomial are related to the n roots $r_1, r_2, r_3, \ldots, r_n$ by

$$\frac{A_{n-1}}{A_n} = -\sum_{i=1}^{n} r_i = -(r_1 + r_2 + r_3 + \cdots)$$

$$\frac{A_{n-2}}{A_n} = \sum_{i<j} r_i r_j = r_1 r_2 + r_1 r_3 + \cdots + r_2 r_3 + \cdots$$

$$\tag{3.3}$$

$$\frac{A_{n-3}}{A_n} = -\sum_{i<j<k} r_i r_j r_k = -r_1 r_2 r_3 - r_1 r_2 r_4 - \cdots$$

$$\vdots$$

$$\frac{A_0}{A_n} = (-1)^n \sum_{i<j<k<\cdots} r_i r_j r_k \cdots = (-1)^n r_1 r_2 r_3 r_4 \cdots = (-1)^n \prod_{i=1}^{n} r_i$$

Theorem 3-7. If the coefficients A_i of $P_n(x)$ are real and $a + bj$ (where $j^2 = -1$) is a root of polynomial $P_n(x) = 0$, then $a - bj$ is also a root.

Theorem 3-8 (Rule of Signs). The number of positive roots of $P_n(x) = 0$, where the coefficients are real, cannot exceed the variation in sign of the coefficients of the polynomial $P_n(x) = 0$. The number of negative roots cannot exceed the number of variations in sign of the coefficients of $P_n(-x)$. Moreover, in each case, if the number of roots is less than the number of variations in sign, it must be smaller by some multiple of two.

Theorem 3-9. To transform an equation of the nth degree into an equation each of whose roots is the negative of the corresponding root of the original equation, we replace the independent variable x by $-x$ in the equation.

Theorem 3-10. To transform an equation of degree n into an equation, each of whose roots is less by k than the roots of the original equation, we proceed as follows: Divide the polynomial by $(x - k)$ and denote the remainder by R_0; divide the quotient by $(x - k)$, and denote the new remainder by R_1; continue until $n + 1$ remainders are found. Then

$$R_n y^n + R_{n-1} y^{n-1} + R_{n-2} y^{n-2} + \cdots + R_1 y + R_0 = 0$$

is the required equation. This can be accomplished through synthetic division as described below.

SYNTHETIC DIVISION

Consider the polynomial

$$f(x) = P_n(x) = A_n x^n + A_{n-1} x^{n-1} + \cdots + A_1 x + A_0 = 0 \tag{3.4}$$

The following relations can be written using the coefficients of this polynomial

$$P_0(x) = A_n$$

$$P_1(x) = x P_0(x) + A_{n-1}$$

$$P_2(x) = x P_1(x) + A_{n-2} \tag{3.5}$$

$$\vdots$$

$$P_n(x) = x P_{n-1}(x) + A_0$$

Multiplying the first, second, third, . . . equations above by x^n, x^{n-1}, x^{n-2}, ... respectively, and adding all of the resulting relations will give Eq. (3.4). In this operation, the terms cross referenced by arrows in Eq. (3.5) will drop out. Thus the value of polynomial $P_n(x)$ at specific value of $x = x_0$ can be obtained by evaluating $P_0(x_0)$ through $P_n(x_0)$ in the above relations. This can be accomplished by synthetic division and is considered simpler than substituting x_0 in Eq. (3.4) and evaluating powers of x_0. The synthetic division will result in

A_n	A_{n-1}	A_{n-2}	$\cdots$	A_1	A_0	$\underline{/x_0}$
	$x_0 P_0(x_0)$	$x_0 P_1(x_0)$	$\cdots$	$x_0 P_{n-2}(x_0)$	$x_0 P_{n-1}(x_0)$	
$P_0(x_0)$	$P_1(x_0)$	$P_2(x_0)$	$\cdots$	$P_{n-1}(x_0)$	$P_n(x_0)$	

where values of P_0 through P_n are sums of respective columns, for example, $P_2 = x_0 P_1 + A_{n-2}$ and values at the second row are obtained by multiplying P_0, P_1, P_2, ... by x_0.

Example. Determine the value of $f(3)$ for the function

$$f(x) = 4x^5 - 30x^3 + 6x^2 - 9$$

by synthetic division. This results in

4	0	-30	6	0	-9	$\diagup 3$
	12	36	18	72	216	
4	12	6	24	72	207	$= f(3)$

DERIVATIVE OF POLYNOMIALS BY SYNTHETIC DIVISION

Differentiating Eqs. (3.5) with respect to x, we have

$$P_0'(x) = 0$$
$$P_1'(x) = P_0(x) + xP_0'(x) = P_0(x) = A_n$$
$$P_2'(x) = xP_1'(x) + P_1(x)$$
$$\vdots$$
$$P_n'(x) = xP_{n-1}'(x) + P_{n-1}(x)$$

(3.6)

where primes denote derivatives with respect to x.

Further differentiation of the above relations will result in

$$P_0''(x) = 0$$
$$P_1''(x) = 0$$
$$P_2''(x) = xP_1''(x) + 2P_1'(x) = 2A_n$$
$$\vdots$$
$$P_n''(x) = xP_{n-1}''(x) + 2P_{n-1}'(x)$$

or

$$\frac{P''_i(x)}{2} = x\left[\frac{P''_{i-1}(x)}{2}\right] + P'_{i-1}(x)$$

where $i = 1, 2, 3, \ldots$.

Further differentiation of the above relations will result in the values of third- and higher-order derivatives

$$\frac{P'''(x)}{3!} \quad \frac{P^{iv}(x)}{4!} \ldots$$

These derivatives can be obtained by extending the synthetic division previously described:

A_n	A_{n-1}	A_{n-2} $\cdots$	A_3	A_2	A_1	$A_0 \;/ x_0$
	$x_0 P_0$	$x_0 P_1$ $\cdots$	$x_0 P_{n-4}$	$x_0 P_{n-3}$	$x_0 P_{n-2}$	$x_0 P_{n-1}$
P_0	P_1	P_2 $\cdots$	P_{n-3}	P_{n-2}	P_{n-1}	$P_n(x_0)$
	$x_0 P'_1$	$x_0 P'_2$ $\cdots$	$x_0 P'_{n-3}$	$x_0 P'_{n-2}$	$x_0 P'_{n-1}$	
P'_1	P'_2	P'_3 $\cdots$	P'_{n-2}	P'_{n-1}	$P'_n(x_0)$	
	$x_0\dfrac{P''_2}{2!}$	$x_0\dfrac{P''_3}{2!}$ $\cdots$	$x_0\dfrac{P''_{n-2}}{2!}$	$x_0\dfrac{P''_{n-1}}{2!}$		
$\dfrac{P''_2}{2!}$	$\dfrac{P''_3}{2!}$	$\dfrac{P''_4}{2!}$ $\cdots$	$\dfrac{P''_{n-1}}{2!}$	$\dfrac{P''_n(x_0)}{2!}$		
	$x_0\dfrac{P'''_3}{3!}$	$x_0\dfrac{P'''_4}{3!}$ $\cdots$	$x_0\dfrac{P'''_{n-1}}{3!}$			
$\dfrac{P'''_3}{3!}$	$\dfrac{P'''_4}{3!}$	$\dfrac{P'''_5}{3!}$ $\cdots$	$\dfrac{P'''_n(x_0)}{3!}$			

These results, together with Taylor's series expansion of $P(x)$ around h

$$P(x + h) = P(h) + xP'(h) + \frac{x^2}{2!}P''(h) + \frac{x^3}{3!}P'''(h) + \dots \qquad (3.7)$$

where h is an incremental quantity, can be used to obtain polynomials whose roots differ from the roots of a given polynomial by the constant factor h. This in turn forms the basis of Horner's method of finding real roots of polynomial expressions to be described shortly.

Example. Given the example

$$P(x) = x^3 - 12x^2 + 47x - 60 = 0$$

write an equation with roots diminished by 2. If the roots of the above equation are $r_1, r_2, \dots$, the roots of

$$P(x + h) = x^3 + B(h)x^2 + C(h)x + D(h) = 0$$

where $B(h)$, $C(h)$, and $D(h)$ are functions of h, will be $(r_1 - h)$, $(r_2 - h)$, $\dots$ since $P(r_i - h + h) = P(r_i) = 0$. Thus, we need only to obtain the coefficients of the Taylor's series (3.7) by synthetic division. This gives

$$
\begin{array}{rrrrl}
1 & -12 & 47 & -60 & \underline{2} \\
 & 2 & -20 & 54 & \\
\hline
1 & -10 & 27 & -6 & = P(2) \\
 & 2 & -16 & & \\
\hline
1 & -8 & 11 & & = P'(2) \\
 & 2 & & & \\
\hline
1 & -6 & & & = P''(2)/2! \\
1 & & & & = P'''(2)/3! \\
\end{array}
$$

Having this, the desired equation becomes

$$P(x + 2) = x^3 - 6x^2 + 11x - 6 = 0$$

HORNER'S METHOD OF FINDING REAL
ROOTS OF POLYNOMIALS

This method involves finding the approximate location of the real root of a polynomial and successively obtaining polynomials whose roots are diminished by these approximate amounts. This method was primarily devised for paper and pencil calculations. For calculators and computers it is much longer than other methods that will be discussed in this chapter. The method is best illustrated by an example. It is desired to find the roots of the equation

$$f(x) = x^3 + 9x^2 - 20x - 90 = 0 \tag{3.8}$$

To obtain the approximate location of a root, set $x = 0, 1, 2, \ldots$. This will result in $f(3) = -42 < 0$ and $f(4) = 38 > 0$. Thus, by Theorem 3-5, $f(x)$ has a root somewhere in the region $3 < x < 4$. The next step is to find a polynomial whose roots are diminished by 3. That is,

$$
\begin{array}{rrrr}
1 & 9 & -20 & -90 \quad \underline{/3} \\
 & 3 & 36 & 48 \\
\hline
1 & 12 & 16 & -42 \\
 & 3 & 45 & \\
\hline
1 & 15 & 61 & \\
 & 3 & & \\
\hline
1 & 18 & &
\end{array}
$$

This results in the polynomial

$$f_1(x) = x^3 + 18x^2 + 61x - 42 = 0 \tag{3.9}$$

The roots of this equation are between 0 and 1. Neglecting higher-order terms, let us take $x = 42/61 \simeq 0.7$ and evaluate $f(0.7)$ by synthetic division:

$$
\begin{array}{rrrr}
1 & 18 & 61 & -42 \quad \underline{/0.7} \\
 & 0.7 & 13.09 & 51.863 \\
\hline
1 & 18.7 & 74.09 & 9.863 = f(0.7)
\end{array}
$$

From this $f(0.7) = 9.863 > 0$, let us try $x = 0.6$; this results in $f(0.6) = 1.296$, again greater than zero.   Let us now try $x = 0.5$; this will give $f(0.5) = -6.875$, which is negative.   Thus, the root of Eq. (3.9) is between 0.5 and 0.6.   Again by synthetic division we write an equation with roots reduced by 0.5 from those of Eq. (3.9).   This results in

1	18	61	– 42	$\diagup$ 0.5
	0.5	9.25	35.125	
1	18.5	70.25	–6.875	
	0.5	9.5		
1	19.0	79.75		
	0.5			
1	19.5			

with the equation

$$f_2(x) = x^3 + 19.5x^2 + 79.75x - 6.875 = 0$$

The roots of this equation are between 0 and 0.1.   Repeating the above process will give the final value of the root of Eq. (3.8) to three significant figures as 3.584.

NEWTON'S METHOD OF SOLVING POLYNOMIAL AND TRANSCENDENTAL EQUATIONS

This method of calculating the roots of polynomial and transcendental equations is used in both desk-type calculators and pencil and paper calculations.   It involves successive approximations to the roots.   Newton's method can also be extended to determine the complex roots of a polynomial equation.   However, in this case the approximations should start with complex values.   In this respect the method suffers from very complicated numerical work in obtaining the complex roots of a polynomial equation.   The following is a description of the method for obtaining the real roots of transcendental and polynomial equations.

Suppose that A is a real root of the equation $f(x) = 0$, and that A_1 is an approximation to A.   Also, let us define the difference by h

$$h = A - A_1$$
$$A = A_1 + h \tag{3.10}$$

Taylor's series expansion for $f(x)$ results in

$$f(A) = f(A_1 + h) = f(A_1) + hf'(A_1) + \frac{h^2}{2!}f''(A_1) + \cdots \tag{3.11}$$

But since A is a root of $f(x)$, then $f(A) = 0$; neglecting terms containing h^2 and higher, we have

$$f(A_1) + hf'(A_1) = f(A_1) + (A - A_1)f'(A_1) = 0$$

From this

$$A = A_1 - \frac{f(A_1)}{f'(A_1)}$$

which is the next approximation to A. Continuing, we arrive at a value for the root.

Example. Find the root of the equation $f(x) = 2 \cos x - 3x = 0$. For $x = 0$, $f(0) > 0$; and for $x = \pi/2$, $f(\pi/2) < 0$. For the first approximation, take midway between $\pi/2$ and zero, that is, $A_1 = \pi/4 \simeq 0.70$. This gives

$$A_2 = A_1 - \frac{f(A_1)}{f'(A_1)} = 0.7 - \frac{2 \cos(0.7) - 2.10}{-2 \sin 0.7 - 3} = 0.567$$

The next approximation gives

$$A_3 = A_2 - \frac{f(A_2)}{f'(A_2)} = 0.56358$$

Further approximations can be obtained by repeating this process.

GRAFFE'S METHOD OF OBTAINING
ROOTS OF POLYNOMIAL
EQUATIONS

This method is based on the fact that if the roots of a polynomial equation are *different* from each other, this difference will be exaggerated if the roots are squared. A polynomial each of whose roots is the square of the roots of the original polynomial can be written. Thus, by repeating this process the roots of the resulting polynomials will be widely separated or, mathematically, $r_1 \gg r_2 \gg r_3 \cdots$ where $r_1, r_2, r_3, \ldots$ are the roots of the succeeding polynomials. As we will see, this makes it possible to use approximations to Eq. (3.3) as a basis of finding the roots. The Graffe method has the advantage of yielding the real as well as the complex roots of a polynomial. In addition, it results in all of the roots rather than one root at a time as in the case of previously given methods. As mentioned before the method is directly applicable when the roots are sufficiently different from each other. If the roots are close to each other in magnitude (or complex conjugate), special procedures should be used to obtain their values. These procedures will be outlined later.

Assuming the variation $r_1 \gg r_2 \gg r_3 \gg r_4 \cdots$ in the roots of the polynomial

$$f(x) = A_n x^n + A_{n-1} x^{n-1} + A_{n-2} x^{n-2} + \cdots + A_{n-1} x + A_0 = 0$$

$$(3.12)$$

Taking the dominant parts of Eq. (3.3), we can relate the above coefficients and the roots

$$r_1 = -\frac{A_{n-1}}{A_n} \qquad r_1 r_2 = \frac{A_{n-2}}{A_n} \qquad r_1 r_2 r_3 = -\frac{A_{n-3}}{A_n} \cdots \qquad (3.13)$$

These relations are obtained from Eq. (3.3) by neglecting smaller-order terms. As mentioned before, the roots of any polynomial equation can be separated by squaring. Take as a preliminary example the quadratic equation

$$x^2 - 3x + 2 = 0 \qquad (3.14)$$

with roots $r_1 = 1$, $r_2 = 2$. Squaring this equation will result in

$$(x^2 + 2)^2 = (3x)^2$$

$$x^4 - 5x^2 + 4 = 0$$

Substituting $y = x^2$ in this equation we obtain

$$y^2 - 5y + 4 = 0$$

with roots 1 and 4. Repeated squaring of the equation will give roots of 1 and 16, 1 and 256, and so on. The corresponding values of the roots of the original equation can be found by using relations (3.13). This results in

$$x^2 - 3x + 2 = 0 \qquad r_1^{(1)} = 3 \qquad\qquad r_2^{(1)} = 0.667$$

$$y^2 - 5y + 4 = 0 \qquad r_1^{(2)} = \sqrt{5} = 2.236 \qquad r_2^{(2)} = 0.894$$

$$z^2 - 17z + 16 = 0 \qquad r_1^{(3)} = \sqrt[4]{17} = 2.031 \qquad r_2^{(3)} = \sqrt[4]{\frac{16}{17}}$$

$$= 0.985$$

$$w^2 - 257w + 256 = 0 \qquad r_1^{(4)} = \sqrt[8]{257} = 2.001 \qquad r_1^{(4)} = \sqrt[8]{\frac{256}{257}}$$

$$= 0.995$$

$$(3.15)$$

Thus, the resulting roots obtained by computing the coefficients are close to the roots of the original Eq. (3.14).

The coefficients of Eq. (3.15) can be evaluated systematically as follows. Consider the polynomial equation

$$f(x) = x^n + b_1 x^{n-1} + b_2 x^{n-2} + b_3 x^{n-3} + \cdots + b_{n-1} x + b_n = 0$$

(Note that in this equation the order of subscripts of b's is reversed as compared to Eq. (3.1). For example, the coefficient of the x^{n-1} term in Eq. (3.1) was A_{n-1}, while in the above equation it is b_1. This is done to ease the explanation which follows.) Transposing odd and even terms in

x to the right and left sides of the equal sign, we get

$$\left(x^n + b_2 x^{n-2} + b_4 x^{n-4} + \cdots\right)$$
$$= -\left(b_1 x^{n-1} + b_3 x^{n-3} + b_5 x^{n-5} + \cdots\right)$$

Squaring both sides of this equation, we obtain

$$x^{2n} - x^{2n-2}\left(b_1^2 - 2b_2\right) + x^{2n-4}\left(b_2^2 - 2b_1 b_3 + 2b_4\right) \tag{3.16}$$
$$- x^{2n-6}\left(b_3^2 - 2b_2 b_4 + 2b_1 b_5 - 2b_6\right) + \cdots = 0$$

Substituting $y = -x^2$ in the above results in

$$y^n + y^{n-1}\left(b_1^2 - 2b_2\right) + y^{n-2}\left(b_2^2 - 2b_1 b_3 + 2b_4\right) + \cdots = 0$$

These coefficients can be obtained from the tabulated form

	x	1	b_1	b_2	b_3	b_4	
First row							
Second row		1	b_1^2	b_2^2	b_3^2	b_4^2	...
Third row			$-2b_2$	$-2b_1 b_3$	$-2b_2 b_4$	$-2b_3 b_5$	...
Fourth row				$2b_4$	$2b_1 b_5$	$2b_2 b_6$	...
$\vdots$					$-2b_6$	$-2b_1 b_7$	...
						$2b_8$	...
$y = -x^2$		1	$b_1^2 - 2b_2$	$b_2^2 - 2b_1 b_3 + 2b_4$	$b_3^2 - 2b_2 b_4 + 2b_1 b_5 - 2b_6$	$b_4^2 - 2b_3 b_5 + 2b_2 b_6 - 2b_1 b_7 + 2b_8$	

The coefficients of x in the original polynomial are written on the first row. The squares of these are written on the second row. The elements of the third row are obtained by multiplying the elements of the first row on either side and multiplying the result by −2. The elements of the fourth row are obtained by multiplying the second element of the first row on either side and multiplying the result by 2. Similarly, the elements of the fifth, sixth, and other rows are obtained. The coefficients of the equation whose roots are the square of the roots of the original equation are obtained by summing the columns of the table from second row and below. Using this method for finding the coefficient, the root-squaring method of Graffe can be applied in finding real and complex roots of polynomial equations, as illustrated in the following examples.

Example. Find the real root of the equation

$$x^3 - 35x^2 - 206x + 240 = 0 \tag{3.17}$$

The table of coefficients for this problem can be written

Root	Coeff x^3	x^2	x^1	x^0
r	1	−35	−206	240
		1.225×10^3	4.24×10^4	5.76×10^4
		0.412×10^3	1.68×10^4	0
r^2	1	1.637×10^3	5.92×10^4	5.76×10^4
		2.68×10^6	3.50×10^9	3.32×10^9
		-0.12×10^6	-0.19×10^9	0
r^4	1	2.56×10^6	3.31×10^9	3.32×10^9
		6.55×10^{12}	1.096×10^{19}	1.102×10^{19}
		-0.006×10^{12}	-0.0008×10^{19}	
r^8	1	6.54×10^{12}	1.095×10^{19}	1.102×10^{19}

From the last row we have

$$|r_1| = \sqrt[8]{6.54 \times 10^{12}} = 39.99 \simeq 40$$

$$|r_2| = \sqrt[8]{\frac{1.095 \times 10^{19}}{6.54 \times 10^{12}}} = 5.997 \simeq 6$$

$$|r_3| = \sqrt[8]{\frac{1.102 \times 10^{19}}{1.095 \times 10^{19}}} = 1.0009 \simeq 1$$

The root-squaring method is terminated when the resulting roots are sufficiently separated, that is, when the original condition of the method is fulfilled. Another way of stating this is whether additional squarings will separate the roots further. As seen from the above table the amount of improvement in the value of the roots is indicated by the cross-product terms, that is, -0.006×10^{12} compared to 6.55×10^{12} and -0.0008×10^{19} compared to 1.096×10^{19} in the last iteration. The squarings are terminated when the contributions of cross-product terms are considered negligible as compared to those of regular terms.

The above absolute values can also be computed using logarithms. It can be seen by substitution in Eq. (3.17) that $r_3 = 1.0009$ is a root. From Eq. (3.3) we have

$$r_1 + r_2 + r_3 = 35$$

$$r_1 r_2 + r_1 r_3 + r_2 r_3 = -206$$

$$r_1 r_2 r_3 = -240$$

Since r_3 is $+1$ from the last equation above, r_1 and r_2 must have different signs; and from the first equation r_1 should be $+40$ and r_2 should be -6.

Complex Conjugate Roots

For polynomials with real coefficients the complex roots occur in conjugate pairs. Since the magnitude of a pair of complex conjugate roots is the same, no amount of root squaring will separate these roots. In these cases, the real roots of the polynomial are calculated first and the complex conjugate pair of roots is determined by using the real roots and the products of complex conjugate roots. Assume that the following polynomial is obtained after several root squarings

$$B_n x^n + B_{n-1} x^{n-1} + B_{n-2} x^{n-2} + \cdots + B_0 = 0 \qquad (3.18)$$

Using Eqs. (3.13) the roots of the polynomial can be written

$$r_1 = \frac{B_{n-1}}{B_n}$$

$$r_1 r_2 = \frac{B_{n-2}}{B_n}; \quad r_2 = \frac{B_{n-2}}{B_{n-1}}$$

$$r_1 r_2 r_3 = \frac{B_{n-3}}{B_n}; \quad r_3 = \frac{B_{n-3}}{B_{n-2}}$$

$$r_1 r_2 r_3 r_4 = \frac{B_{n-4}}{B_n}; \quad r_4 = \frac{B_{n-4}}{B_{n-2}} \tag{3.19}$$

Assuming r_1 and r_4 to be real and r_2 and r_3 to be complex-conjugate pairs, the second of the above relations will involve imaginary numbers. In the root-squaring process this fact results in nonvanishing of cross products in the corresponding B_{n-2} term, as will be seen in the following example. In the Graffe method the presence of a complex-conjugate pair of roots is indicated by the nonvanishing of cross-product terms and many times also by frequent change of sign of these terms.

Example with complex roots. Find the roots of the equation

$$x^3 + 94x^2 - 575x + 2{,}500 = 0 \tag{3.20}$$

The table of coefficients for this equation becomes

Root	Coeff x^3	x^2	x^1	x^0
r	1	94	-575	2,500
		8.836×10^3	3.30×10^5	6.25×10^6
		1.15×10^3	-4.700×10^5	0
r^2	1	9.986×10^3	-1.394×10^5	6.25×10^6
		9.972×10^7	1.943×10^{10}	3.906×10^{13}
		0.028×10^7	-12.483×10^{10}	
r^4	1	10.0×10^7	-10.540×10^{10}	3.906×10^{13}

From the last row

$$|r_1| = \sqrt[4]{10^8} = 100$$

Since this is the absolute value of the root, by substitution in Eq. (3.20) we note that r_1 should be negative $r_1 = -100$. The nonvanishing of the cross-product term in B_{n-2} (-12.483×10^{10}) indicates the presence of a complex root and since $(r_1 r_2)^4 = -10.54 \times 10^{10}$, then r_2 must be imaginary. By Theorem 3-7 r_3 must also be imaginary. Assuming that

$$r_2 = u + jv$$
$$j^2 = -1 \qquad\qquad (3.21)$$
$$r_3 = u - jv$$

we obtain

$$r_2 r_3 = u^2 + v^2 = \frac{r_1 r_2 r_3}{r_1} = \sqrt[4]{\frac{3.906 \times 10^{13}}{10^8}} = 25$$

and since

$$r_1 + r_2 + r_3 = -94 \qquad r_2 + r_3 = 6$$

This together with Eq. (3.21) will result in $2u = 6$ or $u = 3$, and from $u^2 + v^2 = 25$, $v = 4$. Finally, the roots of the equation can be written

$$r_1 = -100 \qquad r_2, r_3 = 3 \pm 4j$$

The presence of real roots of equal magnitude will similarly be indicated by nonvanishing of the cross-product terms. In general, the root-squaring method is unsatisfactory because of special rules it requires in case of multiple roots and complex roots. In the case of all real roots the application of the method is straightforward although the signs of the roots should be determined separately by direct substitution in the equation. In a given problem the real roots can be obtained by, for example, Newton's method. Each root can be factored out of the equation by synthetic division and the process repeated on the polynomial of lower degree. If the polynomial contains only a pair of complex-conjugate roots, the quadratic obtained after extracting all of the real roots can be solved for complex roots. Another often-used method of obtaining the roots of polynomials by digital computers involves successive extraction of quadratic factors by iteration and the solution of the resulting quadratic equations for the roots which may be complex conjugate. The quadratic polynomials can be found by Lin-Bairstow's method, described below.

LIN-BAIRSTOW'S METHOD OF
QUADRATIC FACTORS

Consider the polynomial

$$P_n(x) = A_n x^n + A_{n-1} x^{n-1} + A_{n-2} x^{n-2} + \cdots$$
$$+ A_2 x^2 + A_1 x + A_0 = 0 \tag{3.22}$$

It is desired to extract the quadratic factor $x^2 + r^* x + s^*$ from this polynomial. If $P_n(x)$ is divided by the trial polynomial $x^2 + rx + s$ we obtain a quotient polynomial $Q(x)$ of degree $n-2$ and a remainder $Rx + S$

$$P_n(x) = (x^2 + rx + s)\left(B_n x^{n-2} + B_{n-1} x^{n-3} + B_{n-2} x^{n-4} + \cdots \right.$$
$$\left. + B_3 x + B_2\right) + Rx + S \tag{3.23}$$

Equating the coefficients of the like powers of x in Eqs. (3.22) and (3.23), we get

$$A_n = B_n$$
$$A_{n-1} = B_{n-1} + rB_n$$
$$A_{n-2} = B_{n-2} + rB_{n-1} + sB_n \tag{3.24}$$
$$\vdots$$
$$A_1 = R + rB_2 + sB_3$$
$$A_0 = S + sB_2$$

The coefficients B of the factored polynomial and the terms R and S are functions of r and s. Assuming that $x^2 + r^* x + s^*$ is a factor of the polynomial $P_n(x)$, then

$$R(r^*, s^*) = 0 \qquad S(r^*, s^*) = 0 \tag{3.25}$$

To find r^* and s^* we take values of r and s near these values, so that

$$r^* = r + \Delta r$$
$$s^* = s + \Delta s \tag{3.26}$$

Using Taylor series expansion we have

$$R(r^*, s^*) \cong R(r, s) + \Delta r \frac{\partial R}{\partial r} + \Delta s \frac{\partial R}{\partial s} = 0$$

$$S(r^*, s^*) \cong S(r, s) + \Delta r \frac{\partial S}{\partial r} + \Delta s \frac{\partial S}{\partial s} = 0$$

(3.27)

where the derivatives are evaluated at r and s. Equation (3.27) gives two equations in two unknowns Δr and Δs. Use of these values in Eq. (3.26) will give the next approximation to r^* and s^* and the process is repeated from this point.

The above procedure can be mechanized for a digital computer solution as follows. From Eq. (3.22), if we use the last three terms as the first approximation of the quadratic form, we get the value of r and s

$$r = \frac{A_1}{A_2}$$

$$s = \frac{A_0}{A_2}$$

(3.28)

Equation (3.22) can also be divided by x^{n-2} and the resulting second-order equation used for the initial approximation of the quadratic form. This results in the values of r and s

$$r = \frac{A_{n-1}}{A_n}$$

$$s = \frac{A_{n-2}}{A_n}$$

(3.29)

Having the values of r and s, solve Eqs. (3.24) for B's

$$B_n = A_n$$

$$B_{n-1} = A_{n-1} - rB_n$$

$$B_{n-2} = A_{n-2} - rB_{n-1} - sB_n$$

$$\vdots$$

$$B_{n-i} = A_{n-i} - rB_{n-i+1} - sB_{n-i+2}$$

$$\vdots$$

$$B_3 = A_3 - rB_4 - sB_5$$

$$B_2 = A_2 - rB_3 - sB_4$$

(3.30)

From the last of Eqs. (3.24) compute

$$R = A_1 - rB_2 - sB_3$$

$$S = A_0 - sB_2$$

(3.31)

Noting that B's are functions of r and s, the partial derivatives of R and S with respect to these variables can be computed as follows:

$$\frac{\partial R}{\partial r} = -\left(B_2 + r\frac{\partial B_2}{\partial r}\right) - s\frac{\partial B_3}{\partial r}$$

$$\frac{\partial R}{\partial s} = -r\frac{\partial B_2}{\partial s} - \left(B_3 + s\frac{\partial B_3}{\partial s}\right)$$

$$\frac{\partial S}{\partial r} = -s\frac{\partial B_2}{\partial r}$$

$$\frac{\partial S}{\partial s} = -\left(B_2 + s\frac{\partial B_2}{\partial s}\right)$$

(3.32)

The above equations involve the partial derivatives of B's with respect to r and s. Denoting these partial derivatives by C's, we have, from Eqs. (3.30)*

---

*Note that A's are not functions of r and s.

$$C_n = \frac{\partial B_n}{\partial r} = 0$$

$$C_{n-1} = \frac{\partial B_{n-1}}{\partial r} = -\left(B_n + r\frac{\partial B_n}{\partial r}\right) = -B_n$$

$$\vdots$$

$$C_{n-i} = \frac{\partial B_{n-i}}{\partial r} = -\left(B_{n-i+1} + r\frac{\partial B_{n-i+1}}{\partial r}\right) - s\frac{\partial B_{n-i+2}}{\partial r}$$

$$= -(B_{n-i+1} + rC_{n-i+1}) - sC_{n-i+2}$$

$$\vdots$$

$$C_2 = -(B_3 + rC_3) - sC_4$$

(3.33)

Partial derivatives of B's with respect to s, D's, can also be obtained in a similar fashion

$$D_n = \frac{\partial B_n}{\partial s} = 0$$

$$D_{n-1} = \frac{\partial B_{n-1}}{\partial s} = -r\frac{\partial B_n}{\partial s} = 0$$

$$\vdots$$

$$D_{n-i} = \frac{\partial B_{n-i}}{\partial s} = -r\frac{\partial B_{n-i+1}}{\partial s} - \left(B_{n-i+2} + s\frac{\partial B_{n-i+2}}{\partial s}\right)$$

$$= -rD_{n-i+1} - (B_{n-i+2} + sD_{n-i+2})$$

$$\vdots$$

$$D_2 = \frac{\partial B_2}{\partial s} = -r\frac{\partial B_3}{\partial s} - \left(B_4 + s\frac{\partial B_4}{\partial s}\right) = -rD_3 - (B_4 + sD_4)$$

(3.34)

Substituting these in Eqs. (3.27) and using Eq. (3.32), we get

$$\Delta r\left[-(B_2 + rC_2) - sC_3\right] + \Delta s\left[-(B_3 + sD_3) - rD_2\right] = -R$$

$$\Delta r\left[-sC_2\right] + \Delta s\left[-(B_2 + sD_2)\right] = -S$$

(3.35)

Denoting the coefficients of Δr and Δs by T, U, V, and W in the above equations, we get

$$\Delta r(T) + \Delta s(U) = -R$$
$$\Delta r(V) + \Delta s(W) = -S$$

$$(3.36)$$

Solving these simultaneous equations for Δr and Δs, we have

$$\Delta r = \frac{-RW + SU}{TW - UV}$$
$$\Delta s = \frac{-TS + VR}{TW - UV}$$

$$(3.37)$$

These corrections are applied to the assumed values of r and s and the process is repeated from relations (3.30). The iterations are stopped when sufficient accuracy is achieved between successive values of r and/or s. Having the values of r^* and s^*, the roots are calculated using the relation

$$\text{Roots} = \frac{-r^* \pm \sqrt{(r^*)(r^*) - 4s^*}}{2}$$

$$(3.38)$$

Since with these values of r^* and s^* the remainder terms R and S of Eq. (3.23) are zero, the values of B's will represent the coefficients of a $(n - 2)$-degree polynomial. The roots of this polynomial can be calculated by repeating the above procedure.

TYPICAL COMPUTER LOGIC FOR THE MECHANIZATION OF LIN-BAIRSTOW METHOD

The above equations can be written in flow chart from and subsequently programmed for digital computer solution. The following flow chart describes the logical evaluation of the given equations. The computer program is written in BASIC [4] programming language for a GE 235 time-sharing computer.

The following computer program mechanizes the logic given in the flow chart. The statement numbers corresponding to the blocks of the

Flow Chart of Lin-Bairstow Method

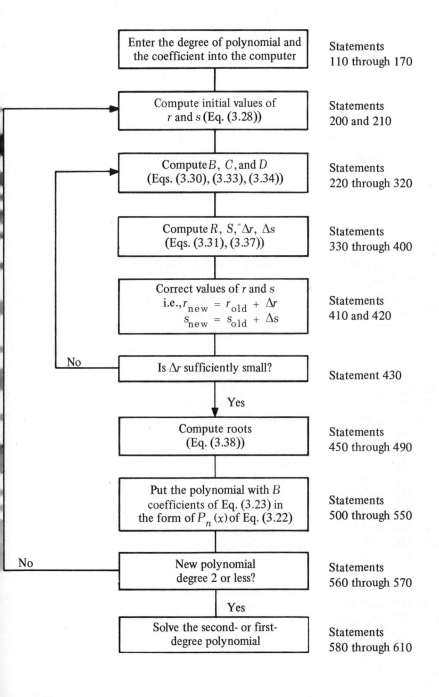

Enter the degree of polynomial and the coefficient into the computer	Statements 110 through 170
Compute initial values of r and s (Eq. (3.28))	Statements 200 and 210
Compute B, C, and D (Eqs. (3.30), (3.33), (3.34))	Statements 220 through 320
Compute R, S, Δr, Δs (Eqs. (3.31), (3.37))	Statements 330 through 400
Correct values of r and s i.e., $r_{new} = r_{old} + \Delta r$ $s_{new} = s_{old} + \Delta s$	Statements 410 and 420
Is Δr sufficiently small?	Statement 430
Compute roots (Eq. (3.38))	Statements 450 through 490
Put the polynomial with B coefficients of Eq. (3.23) in the form of $P_n(x)$ of Eq. (3.22)	Statements 500 through 550
New polynomial degree 2 or less?	Statements 560 through 570
Solve the second- or first-degree polynomial	Statements 580 through 610

No

Yes

No

Yes

flow chart are given next to each block.   The roots of the following polynomial were calculated with this digital computer program

$$x^5 + 3x^4 - 4x^3 - 28x^2 + 43x + 65 = 0 \qquad (3.39)$$

The roots as given at the end of the program are

$$2 \pm j \quad -3 \pm 2j$$

and -1, where $j^2 = -1$. Some of the notations of the computer program are different from those used in the equations.   These are

Equation notation	r	s	R	S	Δr	Δs
Computer notation	R	S	$R1$	$S1$	$R2$	$S2$

From the above discussion it can be concluded that the Lin-Bairstow method of computing the roots of a polynomial equation is adaptable to digital computer programming logic.   It is also seen that this method gives complex roots of polynomials without having to deal with complex numbers during the course of calculation.

COMPUTER PROGRAM FOR LIN-BAIRSTOW'S METHOD

```
100   DIM A(10),B(10),C(10),D(10)
110   READ N
120   PRINT "DEGREE" N
130   PRINT "COEFFICIENT"
140   FOR I = N TO 0 STEP −1
150   READ A(I)
160   PRINT A(I);
170   NEXT I
180   PRINT
190   PRINT
200   LET R=A(1)/A(2)
210   LET S=A(0)/A(2)
220   LET B(N)=A(N)
230   LET C(N)=0
240   LET D(N)=0
250   LET B(N-1)=A(N-1)-R*B(N)
```

```
160   LET C(N-1)= −B(N)
170   LET D(N-1)= 0
180   FOR I=2 TO N−2
190   LET B(N-I)=A(N-I)-R*B(N-I+1)-S*B(N-I+2)
200   LET C(N-I)= -B(N-I+1)-R*C(N-I+1)-S*C(N-I+2)
210   LET D(N-I)= -B(N-I+2)-S*D(N-I+2)-R*D(N-I+1)
220   NEXT I
230   LET R1=A(1)-R*B(2)-S*B(3)
240   LET S1=A(0)-S*B(2)
250   LET T=-B(2)-R*C(2)-S*C(3)
260   LET U= -B(3)-S*D(3)-R*D(2)
270   LET V= -S*C(2)
280   LET W= -B(2)-S*D(2)
290   LET R2 = (-R1*W+S1*U)/(T*W-U*V)
300   LET S2= (-T*S1+V*R1)/(T*W-U*V)
310   LET S=S+S2
320   LET R=R+R2
330   IF ABS (R2) <.000001 THEN 450
340   GO TO 220
350   LET G=R*R-4*S
360   IF G<0 THEN 490
370   PRINT "ROOTS"-R/2"+OR-"SQR(G)/2
380   GO TO 500
390   PRINT "ROOTS"-R/2" +OR-"SQR(-G)/2"J"
400   LET N=N-2
410   PRINT
420   IF N=0 THEN 630
430   FOR I=N TO 0 STEP−1
440   LET A(I)=B(I+2)
450   NEXT I
460   IF N>2 THEN 200
470   IF N<2 THEN 610
480   LET R=A (N-1)/A(N)
490   LET S=A (N-2)/A(N)
500   GO TO 450
510   PRINT "ROOT"-A (N-1)/A(N)
520   DATA 5,1,3,-4,-28,43,65
530   END
```

DEGREE 5
COEFFICIENT

1	3	-4	-28	43	65
ROOTS 2		+OR - 1.	J		
ROOTS-3		+OR - 2.	J		
ROOT −1.					

LAGRANGE'S INTERPOLATION FORMULA
AND INVERSE INTERPOLATION

Assume that $y_0, y_1, y_2, \ldots, y_n$ are values of $y = P(x)$, an unknown polynomial, at $x_0, x_1, x_2, \ldots, x_n$ where the intervals between the x $(x_{i+1} - x_i)$ are not necessarily equal. The problem is to find the polynomial $P(x)$ which passes through $(x_0, y_0), (x_1, y_1), (x_2, y_2), \ldots, (x_n, y_n)$. Having this polynomial, the values of y for any other x can be obtained. This polynomial is called Lagrange's interpolation formula.

Let $A_s(x)$ be a polynomial which is zero for all x's, except x_s, for which it is unity, that is,

$$A_s(x_k) = \begin{cases} 0 & k \neq s \\ 1 & k = s \end{cases}$$

This polynomial can be written in factored form as follows:

$$A_s(x) = A_s(x - x_0)(x - x_1) \cdots (x - x_{s-1})(x - x_{s+1}) \cdots (x - x_n)$$

$$(3.40)$$

Note that the $(x - x_s)$ term is missing on the right of this equation. Setting the value of $A_s(x_s) = 1$ in the above, we can solve for A_s

$$A_s = \frac{1}{(x_s - x_0)(x_s - x_1) \cdots (x_s - x_{s-1})(x_s - x_{s+1}) \cdots (x_s - x_n)}$$

Substituting this expression for A_s in Eq. (3.40), we get

$$A_s(x) = \frac{(x - x_0)(x - x_1) \cdots (x - x_{s-1})(x - x_{s+1}) \cdots (x - x_n)}{(x_s - x_0)(x_s - x_1) \cdots (x_s - x_{s-1})(x_s - x_{s+1}) \cdots (x_s - x_n)}$$

From this we can obtain $A_0, A_1, A_2, A_3, \ldots$ by setting $s = 0, 1, 2, \ldots$. This results in

$$A_0(x) = \frac{(x - x_1)(x - x_2) \cdots (x - x_n)}{(x_0 - x_1)(x_0 - x_2) \cdots (x_0 - x_n)}$$

$$A_1(x) = \frac{(x - x_0)(x - x_2) \cdots (x - x_n)}{(x_1 - x_0)(x_1 - x_2) \cdots (x_1 - x_n)}$$

$$\vdots$$

$$A_i(x) = \frac{(x - x_0)(x - x_1) \cdots (x - x_{i-1})(x - x_{i+1}) \cdots (x - x_n)}{(x_i - x_0)(x_i - x_1) \cdots (x_i - x_{i-1})(x_i - x_{i+1}) \cdots (x_i - x_n)}$$

$$\vdots$$

$$A_n(x) = \frac{(x - x_0)(x - x_1) \cdots (x - x_{n-1})}{(x_n - x_0)(x_n - x_1) \cdots (x_n - x_{n-1})}$$

(3.41)

Note that in the ith equation ($i = 1, 2, 3, \ldots$) the $(x - x_i)$ term is missing in the numerator. Now we form the polynomial $P(x)$ using the coefficients given in Eq. (3.41)

$$P(x) = \sum_{s=0}^{n} A_s(x) y_s = A_0(x) y_0 + A_1(x) y_1 + \cdots + A_n(x) y_n$$

(3.42)

Using the relation $A_s = 0$ for $s \neq k$ and $A_s = 1$ for $s = k$, we get

$$P(x_k) = \sum_{s=0}^{n} A_s(x_k) y_k = A_k(x_k) y_k = y_k$$

which is the desired relation for the value of the polynomial at x_k. Thus, $P(x)$ passes through all points (x_k, y_k) where $k = 0, 1, 2, \ldots$ and the polynomial given by Eq. (3.42) is the desired Lagrange interpolation formula for unequal intervals in x.

The coefficients given by Eq. (3.41) are considerably simplified for equal intervals of the variable x. To obtain these relations consider the following equations:

$$x_s = x_0 + sh$$

$$x = x_0 + uh \tag{3.43}$$

where s is an integer, u can be a fraction, and h is the interval between two consecutive x's. Using these relations the coefficients of the polynomial in Eq. (3.42) become

$$A_0(x) = (-1)^n \frac{(u-1)(u-2) \cdots (u-n)}{n!}$$

$$A_1(x) = (-1)^{n-1} \frac{u(u-2)(u-3) \cdots (u-n)}{1!(n-1)!}$$

$$\vdots$$

$$A_i(x) = (-1)^{n-i} \frac{u(u-1)(u-2) \cdots [u-(i-1)][u-(i+1)] \cdots (u-n)}{i!(n-i)!}$$

$$\vdots$$

$$A_n(x) = \frac{u(u-1)(u-2) \cdots [u-(n-1)]}{n!} \tag{3.44}$$

Note that in the above relations s and u refer to the position of x_s and x with respect to x_0 as given by Eq. (3.43).

Example. As an example of the application of Lagrange's interpolation formulas, consider the evaluation $\sin x$ for $x = 1.2$ radians, given the following values

Interval h	x radians	$\sin x$
0.10	$x_0 = 1.00$	$y_0 = 0.84147$
	$x_1 = 1.10$	$y_1 = 0.89121$
0.20	$x = 1.20$	$y = ?$
0.10	$x_2 = 1.30$	$y_2 = 0.96356$
	$x_3 = 1.40$	$y_3 = 0.98545$

Using the interpolation equations (3.41) for unequal intervals, we have

$$A_0(x) = \frac{(1.2 - 1.1)(1.2 - 1.3)(1.2 - 1.4)}{(1.0 - 1.1)(1.0 - 1.3)(1.0 - 1.4)} = -\frac{1}{6}$$

$$A_1(x) = \frac{(1.2 - 1.0)(1.2 - 1.3)(1.2 - 1.4)}{(1.1 - 1.0)(1.1 - 1.3)(1.1 - 1.4)} = \frac{2}{3}$$

$$A_2(x) = \frac{(1.2 - 1.0)(1.2 - 1.1)(1.2 - 1.4)}{(1.3 - 1.0)(1.3 - 1.1)(1.3 - 1.4)} = \frac{2}{3}$$

$$A_3(x) = \frac{(1.2 - 1.0)(1.2 - 1.1)(1.2 - 1.3)}{(1.4 - 1.0)(1.4 - 1.1)(1.4 - 1.3)} = -\frac{1}{6}$$

Using these values in Eq. (3.42) we get

$$y = \sin x \big|_{\text{at } x = 1.2}$$

$$= -\frac{1}{6}(0.84147) + \frac{2}{3}(0.89121) + \frac{2}{3}(0.96356) - \frac{1}{6}(0.98545)$$

$$= 0.93203$$

The corresponding value of y given in the trigonometric table is $y = 0.93204$.

Inverse Interpolation

The Lagrange interpolation formula can be used for inverse interpolation where the values of $y_0, y_1, y_2, \ldots, y_n$ and the corresponding values of $x_0, x_1, x_2, \ldots, x_n$ are given and it is desired to find the value of x for a specified value of y. Since in most of these cases the values of y are not in equal intervals, Lagrange's formula for unequal intervals is used. The procedure is the same as outlined previously. Note that the Gregory-Newton and Sterling formulas given in Chapter 5 can also be used for interpolation and inverse interpolation.

METHOD OF LEAST SQUARES

The method of least squares is used to find the closest (in a least-squares sense) polynomial approximation to a given set of data. Assume that a set of data consisting of $m + 1$ points

$$(x_0, y_0), (x_1, y_1), (x_2, y_2), \ldots, (x_m, y_m) \tag{3.45}$$

is given. We assume a polynomial of degree $n < m$

$$y = P(x) = A_n x^n + A_{n-1} x^{n-1} + \cdots + A_1 x + A_0 \tag{3.46}$$

and evaluate the coefficients A_n through A_0 by a least-square fit. For this we write the expression for the sum of the squares of the errors in y evaluated by using the assumed polynomial and the given data. That is,

$$\begin{aligned}
\text{Error} = E = &\left[y_0 - \left(A_n x_0{}^n + A_{n-1} x_0{}^{n-1} + \cdots + A_1 x_0 + A_0 \right) \right]^2 \\
&+ \left[y_1 - \left(A_n x_1{}^n + A_{n-1} x_1{}^{n-1} + \cdots + A_1 x_1 + A_0 \right) \right]^2 + \cdots \\
&+ \left[y_n - \left(A_n x_m{}^n + A_{m-1} x_m{}^{n-1} + \cdots + A_1 x_m + A_0 \right) \right]^2
\end{aligned} \tag{3.47}$$

The values of A_n through A_0 can be obtained by minimizing the value of error E with respect to the coefficients $A_n, A_{n-1}, \ldots, A_0$. This is done by setting the partial derivatives of E with respect to these coefficients equal to zero

$$\frac{\partial E}{\partial A_n} = 0$$

$$\frac{\partial E}{\partial A_{n-1}} = 0$$

$$\vdots \tag{3.48}$$

$$\frac{\partial E}{\partial A_0} = 0$$

This will result in $n + 1$ equations in $n + 1$ unknowns A_n through A_0. Solution of these equations will give the coefficients of the assumed polynomial $P(x)$.

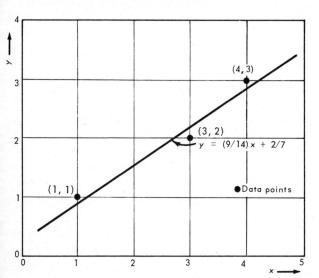

figure 3-1. *Least squares fit to data points.*

Example. Consider data points (1,1), (3,2), (4,3) to be given in a cartesian coordinate system as in Fig. 3-1. It is desired to least-square-fit these points by a straight line

$$y = Ax + B$$

The expression for the sum square error becomes

$$\text{Error} = E = [1 - (A + B)]^2 + [2 - (3A + B)]^2 + [3 - (4A + B)]^2$$

The partial derivatives of E with respect to A and B will result

$$\frac{\partial E}{\partial A} = -2(1 - A - B) - 6(2 - 3A - B) - 8(3 - 4A - B) = 0$$

$$= -19 + 26A + 8B = 0$$

$$\frac{\partial E}{\partial B} = -2(1 - A - B) - 2(2 - 3A - B) - 2(3 - 4A - B) = 0$$

$$= -6 + 8A + 3B = 0$$

or

$$26A + 8B = 19$$
$$8A + 3B = 6$$

Solving the above equation for A and B, we obtain the coefficients of the required polynomial

$$y = \left(\tfrac{9}{14}\right)x + \left(\tfrac{2}{7}\right)$$

Given the data points expressed by Eq. (3.45) and the form of the polynomial (3.46), a general expression can be derived for the coefficients of the set of simultaneous equations with unknown $A_n - A_0$. The simultaneous equations are obtained by writing the expressions for the derivatives of Eq. (3.48) and collecting terms. For example, the partial derivative $\partial E/\partial A_n$ will result

$$
\begin{aligned}
\frac{\partial E}{\partial A_n} = \Big[& 2\left(A_n x_0{}^n + A_{n-1} x_0{}^{n-1} + \cdots + A_0 - y_0\right)x_0{}^n \\
& + 2\left(A_n x_1{}^n + A_{n-1} x_1{}^{n-1} + \cdots + A_0 - y_1\right)x_1{}^n + \cdots \\
& + 2\left(A_n x_m{}^n + A_{n-1} x_m{}^{n-1} + \cdots + A_0 - y_m\right)x_m{}^n \Big] = 0
\end{aligned}
\tag{3.49}
$$

Gathering terms we obtain the first equation in the set of simultaneous equations for $A_n - A_0$,

$$
\begin{aligned}
A_n & \left[x_0{}^{2n} + x_1{}^{2n} + \cdots + x_m{}^{2n}\right] \\
& + A_{n-1}\left[x_0{}^{2n-1} + x_1{}^{2n-1} + \cdots + x_m{}^{2n-1}\right] \\
& + \cdots + A_0\left[x_0{}^n + x_1{}^n + \cdots + x_m{}^n\right] \\
& = y_0 x_0{}^n + y_1 x_1{}^n + \cdots + y_m x_m{}^n
\end{aligned}
\tag{3.50}
$$

The rest of equations can be obtained in a similar manner by setting the derivatives $\partial E/\partial A_{n-1} \cdots \partial E/\partial A_0$ of Eq. (3.48) equal to zero. The set of simultaneous equations for $A_m - A_0$ can then be written

$$A_n[2n] + A_{n-1}[2n - 1] + \cdots + A_0[n] = [y(n)]$$
$$A_n[2n - 1] + A_{n-1}[2n - 2] + \cdots + A_0[n - 1] = [y(n - 1)] \quad (3.51)$$
$$\vdots$$
$$A_n[n] + A_{n-1}[n - 1] + \cdots + A_0[0] = [y(0)]$$

where the notation

$$\sum_{i=0}^{m} x_i^{\,j} = [j] \qquad (3.52)$$

and

$$\sum_{i=0}^{m} y_i\, x_i^{\,j} = [y(j)]$$

is used.   Denoting the matrix of the coefficients of A's by the square matrix S and the right side of Eq. (3.51) by the column matrix T, we get

$$SA = T \qquad (3.53)$$

where A denotes the column matrix of the coefficients $A_m - A_0$. This matrix can be obtained by premultiplying Eq. (3.53) by the inverse of the S matrix, that is,

$$A = S^{-1}T$$

The inverse matrix can be obtained, for example, by the augmented matrix method described in Chap. 1.

COMPUTER PROGRAM FOR LEAST-SQUARES METHOD

The following computer program has been written in FORTRAN (see [5]) programming language to implement the calculation of the coefficients in the least-squares polynomial fit. Statements 100 and 110 reserve computer storage space for the following variables.

Program Symbol	Test Symbol	Description
X	x	Data points
Y	y	Data points
C	[j]	Eq. (3.52)
B	[y(j)]	Eq. (3.52)
S	S	Matrix of the coefficients, Eq. (3.53)
T	T	Eq. (3.53)
A	A	Matrix of the coefficients, $A_n - A_0$
CY		The values of the variable y computed from the fitted polynomial
N		Degree of the polynomial
M		The number of data points

Statement 120 reads the degree of the polynomial to be fitted N and the number of data points M. Statement 130 decreases the number of data points by one to compensate for starting from $x_0, y_0; x_1, y_1; \cdots$ and so on. Statement 140 reads the coordinates of the data points to be fitted. Statements 150-250 compute the values of the coefficients given by Eq. (3.52), C's and B's of the program. Note that the symbol ↑ specifies power. Statements 260-340 set these values in the matrix form of the S matrix of the coefficients. Note here that since the elements of the matrix start from the first row and the first column, the number N representing the degree of the polynomial is incremented by unity in statement 260. Statements 350 through 650 are exactly the same as the one given in Chap. 1 for augmented matrix inversion method. At statement 660 the inverse of the S matrix will be calculated and available. Referring to the augmented matrix method and subsequent program of Chap. 1, we see that the inverse matrix is stored in columns $NX = M + 1$ and $NY = 2N$ rather than 1 to N. This fact is used in statements 660 through 700 in calculating the product $S^{-1}T$ which represents the coefficients $A_n - A_0$. Statement 710 prints these coefficients.

```
100   DIMENSION  X(25),Y(25),C(25),B(25),S(25,25),T(25),A(25)
110   DIMENSION  CY(25)
120   READ, N,M
130   M=M-1
140   READ,  (X(I),Y(I),I=0,M)
150   NY=2*N
160   DO  10  I=0,NY
170   C(I)=0.
180   DO 10 J=0,M
190   C(I)=C(I)+X(J)↑I
200   10:
210   DO 20 I=0,N
220   B(I)=0.
230   DO 20 J=0,M
240   B(I)=B(I)+Y(J)*(X(J) ↑ I)
250   20:
260   N=N+1
270   DO 30 J=1,N
280   DO 30 I=1,N
290   NI=2*N-I-J
300   S(I,J)=C(N1)
310   30:
320   DO 35 I=1,N
330   T(I)=B(N-I)
340   35:
350   I=1
360   NX=N+1
370   NY=2*N
380   DO 40 J=NX,NY
390   S(I,J)=1.
400   I=I+1
410   40:
420   L=1
430   K=2
440   42: XM=S(L,L)
450   DO 45 J=L,NY
460   S(L,J)=S(L,J)/XM
470   45:
480   DO 50 I=K,N
490   XZ=S(I,L)
500   DO 50 J=L,NY
510   S(I,J)=S(I,J)-S(L,J)*XZ
520   50:
530   L=L+1
540   K=K+1
```

```
550   IF (L-N) 42,42,52
560   52: L=N
570   53: LZ=L-1
580   DO 60 K=1, LZ
590   I=L-K
600   YZ=S(I,L)
610   DO 60 J=L,NY
620   S(I,J)=S(I,J)-S(L,J)*YZ
630   60:
640   L=L-1
650   IF(L-I)65,65,53
660   65: DO 70 I=1,N
670   A(I)=0.
680   DO 70 J=NX,NY
690   A(I)=S(I,J)*T(J-N)+A(I)
700   70:
710   PRINT,(A(I),I=1,N)
720   PRINT, ↑↑↑↑↑↑
730   DO 80 J=0,M
740   CY (J)=0.
750   DO 75 I=1,N
760   CY(J)=CY(J)+A(I)*(X(J) ↑ (N-I))
770   75:
780   80: PRINT,X(J),Y(J),CY(J),CY(J)-Y(J)
790   $DATA
800   2,3
810   1,1,3,2,4,3
820   END
```

Statements 730 through 780 calculate the values of y at data points $x_0 - x_m$, using the polynomial coefficients just developed. The results of the computer run for the case of the previously given example are shown below.

Coefficients for a first-degree polynomial

$$A_1 = 0.6429 \quad A_0 = 0.2857$$

x	y_{data}	y_{comp}	Error $y_{comp} - y_{data}$
1.00	1.00	0.9286	−0.0714
3.00	2.00	2.2143	0.2143
4.00	3.00	2.8571	−0.1429

The coefficients of the polynomial fit can equally well be computed for a second-degree polynomial. In this case we have

Coefficients for a second-degree polynomial

$$A_2 = 0.1667 \qquad A_1 = -0.1667 \qquad A_0 = 1.00$$

and

x	y_{data}	y_{comp}	Error $y_{comp} - y_{data}$
1.00	1.00	1.00	0.00
3.00	2.00	2.00	0.00
4.00	3.00	3.00	0.00

This result is expected, since a second-degree polynomial will pass through the three data points.

PROBLEMS

1. Determine the number of positive and negative roots of the equation

$$x^5 - 3x^4 + 4x^3 + 3x^2 + 6x + 3 = 0$$

2. Find the value of the function

$$f(x) = x^4 + 6x^3 + 8x + 12$$

   at $x = 3$ and 5 by synthetic division. Check your solution by substitution in $f(x)$.

3. Determine the polynomial each of whose roots is diminished by 2 from the roots of the polynomial

$$x^3 - 8x^2 + 19x + 12 = 0$$

4. Find the root of the polynomial

$$x^3 - 5x^2 + 6x - 1 = 0$$

   between 1 and 2 by Horner's method.   *Answer:* 1.555.

5. Find the root of the equation

   $$\tan x = 1 - x$$

   by Newton's method.

6. Find the roots of the equation given in Prob. 3 by Graffe's method.

7. Find the roots of the equation

   $$x^3 - 5x^2 + 8x - 6 = 0 \qquad Answer: \begin{pmatrix} i - j \\ i + j \\ 3 \end{pmatrix}$$

   by Graffe's method.

8. Find the roots of equation

   $$x^3 - 5x^2 + 6x - 1 = 0 \qquad Answers:\ 3.2469,\ 1.5549,\ .19806$$

   by Graffe's method.

9. Find the roots of the equation

   $$x^3 - 2x^2 - 8x + 5 = 0 \qquad Answer:\ -2.3375,\ 3.7702,\ .5674$$

   by Lin-Bairstow's method.

10. Find the roots of the equation

    $$x^3 + 94x^2 - 575x + 2,500 = 0 \qquad Answer:\ -100,\ 3 \pm 4j$$

    by Lin-Bairstow's method.

11. Find the real roots of the equations of Probs. 9 and 10 by Newton's method.

12. Given the table of values

n	$\sqrt{n}$
150	12.24745
152	12.32883 A
154	12.40967
156	12.49000
158	12.56981 B
160	12.64911

find $\sqrt{155}$ using Lagrange's interpolation formula for equal intervals using the A and B sets of values, respectively.

13. Using the table of values

n	$\sqrt[3]{n}$
50	3.684
52	3.732
54	3.779
56	3.825

find the value of n given $\sqrt[3]{n} = 3.756$ using Lagrange's interpolation formula (inverse interpolation). *Answer:* 53.

14. Given the table of values

x	y
0	6
1	11
2	22
3	39

find the coefficients of the equation $y = Ax^2 + Bx + C$ for a least-square fit to the given points. *Answer:* $y = 3x^2 + 2x + 6$.

15. Find the coefficients of the equation $y = Ax^2 + Bx + C$ for a least-square fit of the table of Prob. 13. Assume that $x = n$ and $y = \sqrt[3]{n}$. *Answer:* $y = -.000125\, x^2 + .03675\, x + 2.159$

16. Verify the terms of Eq. (3.51) for $n = 2$ and $m = 3$.

REFERENCES

1. Sahon, Harry: "Engineering Mathematics," D. Van Nostrand Co., Inc., Princeton, N. J., 1949.
2. Kunz, K. S.: "Numerical Analysis," McGraw-Hill Book Company, New York, 1957.
3. Arley, N., and K. R. Buch: "Introduction to the Theory of Probability and Statistics," John Wiley & Sons, Inc., New York, 1961.
4. BASIC Language, reference manual, General Electric Company, May, 1966.

5. Farina, Mario V.: "Fortran IV," Prentice-Hall, Inc., Englewood Cliffs, N. J., 1966.

4 TIME-FREQUENCY DOMAIN ANALYSES AND THE FAST FOURIER TRANSFORM

In engineering analyses, parameters of system behavior are either expressed as a function of time or as a function of the frequency contents. For example, the solution of differential equations of motion of rigid bodies will result in the values of displacements, velocities, and accelerations as a function of time, while oscillations of electrical and mechanical systems and doppler shifts of radar return are expressed in the frequency domain. Time and frequency domain analyses are related through Fourier series and transforms. Given a variable expressed as a function of time, Fourier analysis will decompose the variable into a sum of oscillatory functions, each having a specific frequency. These frequencies, with their corresponding amplitudes and phase angles, constitute the frequency contents of the original variable.

A newly developed technique of Fourier transform analyses [1] for digital computers, called the fast Fourier transform method, has made the use of these computers in frequency domain analyses very attractive. The method has been available in mathematical literature for some time [2] but it has been rediscovered by Cooley and Tukey only recently [3]. The Cooley-Tukey method reduces the number of multiplications required in calculating the frequency contents of a time function by a factor of γ/N, as compared to the direct method of computation, where $N = 2^\gamma$ is the number of discrete frequencies examined and γ is an integer. The Cooley-Tukey method is primarily based on using binary logic for computation and for this reason the number of discrete frequency steps are taken in powers of two.

In this chapter we will discuss the discrete Fourier analysis using matrices for computations, and we will proceed to describe the Cooley-Tukey computational process. The report also includes a digital computer program for obtaining the frequency contents of time functions.

DISCRETE FOURIER TRANSFORM

The exponential form of Fourier series of the periodic function $F(t)$, with period T, is expressed as

$$F(t) = \sum_{n=-\infty}^{\infty} a_n e^{jn\omega t} \tag{4.1}$$

and

$$a_n = \frac{1}{T} \int_{-T/2}^{T/2} F(t) e^{-jn\omega t} dt \tag{4.2}$$

where

ω = angular frequency = $2\pi f$
f = frequency = $1/T$
T = period
j = $\sqrt{-1}$

In Eqs. (4.1) and (4.2) it is assumed that the function is periodic with period T and the coefficient a_n represents the integral of the function over a complete period. The corresponding Fourier transform pair of the function $F(t)$ is expressed

$$F(t) = \int_{-\infty}^{\infty} G(\omega) e^{j\omega t} d\omega \tag{4.3}$$

and

$$G(\omega) = \frac{1}{2\pi} \int_{-\infty}^{\infty} F(t) e^{-j\omega t} dt \tag{4.4}$$

where $G(\omega)$ is the exponential Fourier transform of $F(t)$. Equation (4.4) of the Fourier transform can be compared to Eq. (4.2) of the Fourier series if one assumes the period of the function $F(t)$ to extend over the entire time domain. In practice, however, the function $F(t)$ is given starting from time zero and extending over a span of time T, as given in Fig. 4-1(a). This will set the limits of Eq. (4.4) between zero and T. Given the Fourier transform $G(\omega)$ of the function $F(t)$, Eq. (4.3), referred to as the inversion integral, will result in the function $F(t)$. For the discrete representation of Eqs. (4.3) and (4.4), consider Fig. 4-1. The time span T of Fig. 4-1(a) is divided into K equal increments of Δt each. For convenience and efficiency of calculations, let us set $K = 2^\gamma$, where γ is an integer. The angular frequency span ω_N is divided into N equal increments of $\Delta\omega$ each. Denote N in the binary form $N = 2^{\gamma \pm \gamma_1}$ where γ_1 is an integer. Note that although this definition allows the values of N and K to be different, in practice they are taken equal. The frequency interval $\Delta\omega$ is defined as

$$\Delta\omega = 2\pi \Delta f = \frac{1}{T(N/K)} = \frac{2\pi K}{NT} \tag{4.5}$$

In the above, the value of Δf is set at K/NT. This is needed to simplify the method of calculation. Equations (4.3) and (4.4) with the above

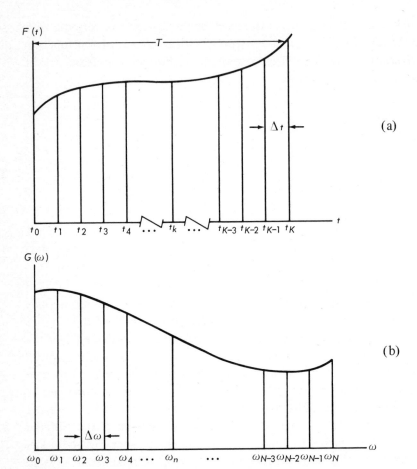

figure 4-1. *Discrete representation of* $F(t)$ *and* $G(\omega)$.

notation can be written in discrete form as follows

$$F(t_k) = \Delta\omega \sum_{n=0}^{N-1} G(\omega_n) e^{j\omega_n t_k} \tag{4.6}$$

$$G(\omega_n) = \frac{\Delta t}{2\pi} \sum_{k=0}^{K-1} F(t_k) e^{-j\omega_n t_k} \tag{4.7}$$

where the limits of the first sum instead of $-N/2$ to $N/2$ (corresponding to Eq. (4.3)) are set from 0 to $N-1$ for computational convenience.

As will be seen later, this changes the foldover frequency from zero to $\omega_{N/2}$. The limits of Eq. (4.7) are taken to represent the available range of the variable $F(t)$. Equations (4.6) and (4.7) represent the areas under the curves of Fig. 4-1. Using the following definitions,

$$F'_k = F(t_k)$$

$$G_n = G(\omega_n)$$

$$\omega_n = n\,\Delta\omega = \frac{2\pi nK}{NT} \tag{4.8}$$

$$t_k = k\,\Delta t = \frac{kT}{k}$$

Eqs. (4.6) and (4.7) can be written

$$F'_k = \Delta\omega \sum_{n=0}^{N-1} G_n e^{(2\pi j/N)(nk)} \tag{4.9}$$

$$G_n = \frac{T}{2\pi K} \sum_{k=0}^{K-1} F'_k e^{(-2\pi j/N)(nk)} \tag{4.10}$$

Denoting $W = e^{-2\pi j/N}$, incorporating the constant $T/2\pi K$ into the function F'_k, and denoting the product as F_k, Eq. (4.10) can be written in matrix form as

$$[G_n] = [W^{(nk)}][F_k] \qquad \begin{matrix} n = 0, 1, 2, \ldots, N-1 \\ k = 0, 1, 2, \ldots, K-1 \end{matrix} \tag{4.11}$$

$$(n \times 1)(n \times k)(k \times 1)$$

where subscripts n and k represent the row numbers of column matrices $[G]$ and $[F]$, respectively, and the product nk in addition to being the power of W represents the row n and column k location of the W^{nk} element. Before discussing the solution of matrix equation (4.11), consider the exponential function

$$W^l = e^{(-2\pi j/N)\, l} = \cos\left(\frac{2\pi}{N}\right)l - j \sin\left(\frac{2\pi}{N}\right)l \tag{4.12}$$

where $l = nk$ is the exponent. Since the exponential function of Eq. (4.12) repeats itself in multiples of N, the values of l can be set

$$
\begin{aligned}
l &= nk \\
&= nk \text{ modulus } N \\
&= \text{integer remainder after division of } nk \text{ by } N
\end{aligned}
\tag{4.13}
$$

Under these conditions, l will assume values $0,1,2,3,\ldots, N$. It is further noted from Eq. (4.12) that

$$W^0 = W^N = 1 \tag{4.14}$$

and

$$W^{N/2} = e^{-\pi j} = -1 = -W^0$$

As an example, let us take the value of $N = 8 = 2^3$ and calculate the values of W^l where $l = 0,1,2,\ldots, 8$. Using the relation

$$W^l = e^{(-2\pi j/8)\, l} = \cos\left(\frac{\pi l}{4}\right) - j \sin\left(\frac{\pi l}{4}\right) \qquad l = 0, 1, \ldots, 8 \tag{4.15}$$

From this, the values of W^l can be calculated and plotted on the complex plane as shown in Fig. 4-2. It is noted that the foldover frequency, where W^i and W^{-i} are symmetrically located about the real axis, occurs at ω_4 corresponding to the angular frequency of $W^{N/2}$ or W^4. The values of $W^l (l > N/2)$ are therefore the complex conjugates of the values of W^{N-l}.

MATRIX CALCULATION OF
FOURIER TRANSFORMS

In order to effectively explain the logic of calculating the values of the matrix equation (4.11), let us set $N = K = 8$ and write the expanded

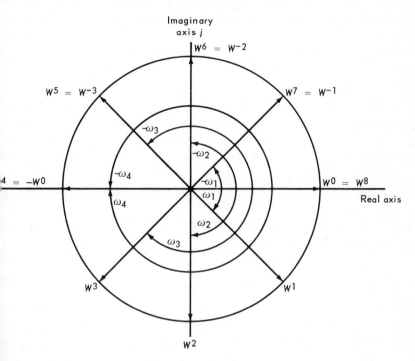

Figure 4-2. *Powers of the exponential function* W.

orm of this equation as follows:

$$
\begin{array}{cccccccccc}
\text{Column} \rightarrow & 0 & 1 & 2 & 3 & 4 & 5 & 6 & 7 & 0
\end{array}
$$

$$
\begin{bmatrix} G_0 \\ G_1 \\ G_2 \\ G_3 \\ G_4 \\ G_5 \\ G_6 \\ G_7 \end{bmatrix}
=
\begin{bmatrix}
W^{0(0,0)} & W^{0(0,1)} & W^{0(0,2)} & W^{0(0,3)} & W^{0(0,4)} & W^{0(0,5)} & W^{0(0,6)} & W^{0(0,7)} \\
W^{0(1,0)} & W^{1(1,1)} & W^{2(1,2)} & W^{3(1,3)} & W^{4(1,4)} & W^{5(1,5)} & W^{6(1,6)} & W^{7(1,7)} \\
W^{0(2,0)} & W^{2(2,1)} & W^{4(2,2)} & W^{6(2,3)} & W^{0(2,4)} & W^{2(2,5)} & W^{4(2,6)} & W^{6(2,7)} \\
W^{0(3,0)} & W^{3(3,1)} & W^{6(3,2)} & W^{1(3,3)} & W^{4(3,4)} & W^{7(3,5)} & W^{2(3,6)} & W^{5(3,7)} \\
W^{0(4,0)} & W^{4(4,1)} & W^{0(4,2)} & W^{4(4,3)} & W^{0(4,4)} & W^{4(4,5)} & W^{0(4,6)} & W^{4(4,7)} \\
W^{0(5,0)} & W^{5(5,1)} & W^{2(5,2)} & W^{7(5,3)} & W^{4(5,4)} & W^{1(5,5)} & W^{6(5,6)} & W^{3(5,7)} \\
W^{0(6,0)} & W^{6(6,1)} & W^{4(6,2)} & W^{2(6,3)} & W^{0(6,4)} & W^{6(6,5)} & W^{4(6,6)} & W^{2(6,7)} \\
W^{0(7,0)} & W^{7(7,1)} & W^{6(7,2)} & W^{5(7,3)} & W^{4(7,4)} & W^{3(7,5)} & W^{2(7,6)} & W^{1(7,7)}
\end{bmatrix}
\begin{bmatrix} F_0 \\ F_1 \\ F_2 \\ F_3 \\ F_4 \\ F_5 \\ F_6 \\ F_7 \end{bmatrix}
$$

$$(4.16)$$

where the power of W, given outside the parentheses, is computed using relation (4.13). The row number n and the column number k are given within the parentheses (n, k). It is also noted that the W matrix is a symmetric matrix and its elements, calculated from Eq. (4.15), are complex numbers.

Using the usual method of matrix multiplication, the G matrix can be calculated by multiplying the rows of the W matrix by the column of the F matrix. This will take N^2 multiplications, or in this case, 64 multiplications. In the following discussion, it will be shown that the number of multiplications can be considerably reduced if one uses some of the properties of the W matrix. The product of W and F matrices can be written as

$$\text{Column} \rightarrow \quad 0 \qquad 1 \qquad 2 \qquad\qquad 7$$

$$G_0 = F_0 + W^0 F_1 + W^0 F_2 + \cdots + W^0 F_7$$

$$G_1 = F_0 + W^1 F_1 + W^2 F_2 + \cdots + W^7 F_7$$

$$G_2 = F_0 + W^2 F_1 + W^4 F_2 + \cdots + W^6 F_7$$

$$G_3 = F_0 + W^3 F_1 + W^6 F_2 + \cdots + W^5 F_7$$

$$G_4 = F_0 + W^4 F_1 + W^0 F_2 + \cdots + W^4 F_7 \qquad\qquad (4.17)$$

$$G_5 = F_0 + W^5 F_1 + W^2 F_2 + \cdots + W^3 F_7$$

$$G_6 = F_0 + W^6 F_1 + W^4 F_2 + \cdots + W^2 F_7$$

$$G_7 = F_0 + W^7 F_1 + W^6 F_2 + \cdots + W^1 F_7$$

where W^0 of the first column of the W matrix of Eq. (4.16) is substituted by unity. The computation of $G_0 G_1 \ldots G_7$ of Eq. (4.17) can be accomplished by initially setting their values equal to F_0, zeroth column, and adding the elements of the first column, and then the second column, and so on. It is noted that all of the W's of the first column are multiplied by F_1, and those of the second column by F_2, and so on. We further observe that in computing the elements $W^0 F_1$, $W^1 F_1$, $W^2 F_1, \ldots$, of the first column, the value of $W^{N/2} = W^4$ is equal to $-W^0 = -1$, and the values of W after $W^{N/2}$ are complex conjugate of previously computed values symmetrically located with respect to $W^{N/2}$. For example, we only have to compute

$$W^0 F_1 \qquad W^1 F_1 \qquad W^2 F_1 \quad \text{and} \quad W^3 F_1$$

nce the rest of the values can be obtained from these, i.e.,

$$W^4 F_1 = -W^0 F_1 \quad W^5 F_1 = \overline{W^3 F_1} \quad W^6 F_1 = \overline{W^2 F_1} \tag{4.18}$$

and $\quad W^7 F_1 = \overline{W^1 F_1}$

where bars denote complex conjugates.

From the second column of Eq. (4.17), it is noted that the values of the products repeat after the second $W^0 F_2$ is encountered. Thus, in this case, we only have to compute the values of $W^0 F_2$ and $W^2 F_2$ since

$$W^4 F_2 = -W^0 F_2 \quad \text{and} \quad W^6 F_2 = \overline{W^2 F_2}$$

and the rest of the values $W^0 F_2$, $W^2 F_2$, $W^4 F_2$, and $W^6 F_2$ are merely a repetition of the previously calculated set of products. The above procedure can be summarized into the following set of computational rules:

1. Denote the number of time increments by K, frequency increments by N, and the total time span by T.

2. Obtain the values of W^l where $l = 0,1,2, \ldots, N$, using Eq. (4.15).

3. Set the values of $G_0, G_1, \ldots, G_{N-1}$ each equal to F_0 (Eq. (4.17), zeroth column).

4. Compute the value of l, nk modulus N, where n is the row number and k is the column number of the elements of the W matrix of Eq. (4.16), starting with $k = 1$ and $n = 0,1,2, \ldots, N-1$. Multiply the corresponding W by F_k. The first l will always be zero as it corresponds to the power of W in the zeroth row of the matrix of Eq. (4.16). When $l = N/2$ obtain the complex conjugates of the previously calculated WF products, symmetrically located with respect to $W^{N/2}$. If $l = 0$ the rest of the values of WF repeat.

5. Add the values computed in step (4) to the existing values of $G_0, G_1, \ldots, G_{N-1}$.

6. Repeat steps (4) and (5) for $k = 2,3,4 \ldots, N-1$.

The logic of the above procedure can be programmed for digital computer solution. Once such program is given following the numerical examples.

The values of G calculated above will be complex numbers. Th element G_0 will correspond to angular frequency $\omega = 0$, G_n to $\omega = n\,\Delta$ and so on. It will be shown in the following discussion that the values G symmetrically located with respect to $G_{N/2}(= G_4$ for $N = 8)$ in th column matrix of Eq. (4.16) will be complex conjugates, that $G_5 = \overline{G_3}$, $G_6 = \overline{G_2}$, and $G_7 = \overline{G_1}$. This fact will eliminate calculation G's beyond $G_{N/2}$.

Consider the values of G_2 and G_6 calculated from matrix equatio (4.16) and given below:

$$G_2 = \underline{W^0 F_0} + W^2 F_1 + \underline{W^4 F_2} + W^6 F_3 + \underline{W^0 F_4}$$

$$+ W^2 F_5 + \underline{W^4 F_6} + W^6 F_7$$

$$G_6 = \underline{W^0 F_0} + W^6 F_1 + \underline{W^4 F_2} + W^2 F_3 + \underline{W^0 F_4}$$

$$+ W^6 F_5 + \underline{W^4 F_6} + W^2 F_7$$

(4.19

Since $W^0 = -W^4 = 1$, the underlined values in the above equation will b real. From Fig. 4-2 it is observed that W^2 and W^6 are complex conjugate Using this fact in Eq. (4.19) will show that G_2 and G_6 are also comple conjugates.

For future discussion, denote

$$G_n = a_n + jb_n$$

where a_n and b_n are the real and imaginary parts of G_n. The trigono metric and the exponential forms of G_n can be written as follows

$$G_n = |G_n| (\cos \psi_n + j \sin \psi_n)$$

$$= |G_n| e^{j\psi_n}$$

where

$$|G_n| = \text{absolute value of } G_n$$

$$= \sqrt{a_n{}^2 + b_n{}^2}$$

$$\psi_n = \text{arc tan } \frac{b_n}{a_n}$$

The value of G_n can be substituted for $G(\omega_n)$ in Eq. (4.6) to obtain the inverse Fourier transform.

Example 4-1

As the first example of obtaining the frequency contents by Fourier transform, consider the wave of Fig. 4-3(a). Setting the time and frequency steps at $K = 16$ and $N = 16$, the F matrix of Eq. (4.16) can be calculated by multiplying $F(t)$ by $T/2\pi K$ at each of the time increments $\Delta t = T/K = 1$. The sixteen values of W^{nk} are then calculated using Eq. (4.12). From Eq. (4.5) the value of $\Delta\omega = 2\pi/16 = 0.39269$. Having these values and using the logic described above, the values of G_0 through G_{15} can be calculated.

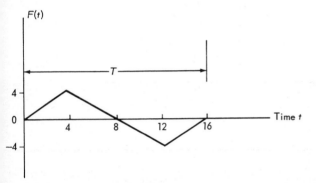

figure 4-3(a). $F(t)$ vs. time, Example 1.

Table 4-1 gives a list of the values calculated by the digital computer program to be described. The values of the function $F(t)$ are shown in the first part of Table 4-1 while their Fourier transforms are shown on the lower part of the table. The fold over occurs at $G_{N/2} = G_8$ corresponding to $\omega = \pi$. It is further observed that the values of G's symmetrically located with respect to G_8 are complex conjugates.

For the purpose of discussion, let us write the Fourier series [4] of the time function of Fig. 4-3(a)

$$F(t) = \frac{32}{\pi^2}\left(\sin\omega t - \frac{1}{3^2}\sin 3\omega t + \frac{1}{5^2}\sin 5\omega t - \cdots\right) \tag{4.20}$$

$$= 3.2423\sin\omega t - 0.3602\sin 3\omega t + 0.1297\sin 5\omega t - \cdots$$

Table 4-1.   Solution of Example Problem 4-1

Time step k	Time t	Function $F(t)$
0	0	0
1	1	1
2	2	2
3	3	3
4	4	4
5	5	3
6	6	2
7	7	1
8	8	0
9	9	−1
10	10	−2
11	11	−3
12	12	−4
13	13	−3
14	14	−2
15	15	−1

Freq. step n	Omega ω	Absolute value of $\lvert G \rvert$	Real part of G	Imaginary part of G
0	0	0	0	0
1	0.392699	4.18166	0	−4.18166
2	0.785398	0	0	0
3	1.1781	0.515635	0	0.515635
4	1.5708	0	0	0
5	1.9635	0.230212	0	−0.230212
6	2.35619	0	0	0
7	2.74889	0.165452	0	0.165452
8	3.14159	0	0	0
9	3.53429	0.165452	0	−0.165452
10	3.92699	0	0	0
11	4.31969	0.230212	0	0.230212
12	4.71239	0	0	0
13	5.10509	0.515635	0	−0.515635
14	5.49779	0	0	0
15	5.89049	4.18166	0	4.18166

where $\omega = 2\pi/T = 0.392699$.   Consider the inversion Eq. (4.6) of the Fourier transform

$$
\begin{aligned}
F(t) \;=\; \Delta\omega \Big\{ &(a_0 + jb_0)\,[\cos\omega_0 t + j\sin\omega_0 t] \\
+\; &(a_1 + jb_1)\,[\cos\omega_1 t + j\sin\omega_1 t] \\
+\; &(a_2 + jb_2)\,[\cos\omega_2 t + j\sin\omega_2 t] \\
+\; &\cdots \Big\}
\end{aligned}
\tag{4.21}
$$

where a_n and b_n, as before, are the real and imaginary parts of G_n. Substituting the values of a and b from Table 4-1, we get

$$
\begin{aligned}
F(t) \;=\; 0.392699 \Big\{ &(-4.18166j)\,(\cos\omega_1 t + j\sin\omega_1 t) \\
+\; &(0.515635j)\,(\cos\omega_3 t + j\sin\omega_3 t) \\
+\; &(-0.230212j)\,(\cos\omega_5 t + j\sin\omega_5 t) \\
+\; &(0.165452j)\,(\cos\omega_7 t + j\sin\omega_7 t) \\
+\; &(-0.165452j)\,(\cos\omega_9 t + j\sin\omega_9 t) \\
+\; &(0.230212j)\,(\cos\omega_{11} t + j\sin\omega_{11} t) \\
+\; &\cdots \Big\}
\end{aligned}
\tag{4.22}
$$

Using the relations $\omega_{15} = -\omega_1$, $\omega_{13} = -\omega_3$, and $\omega_{11} = -\omega_5$, Eq. (4.22) simplifies to the form

$$
F(t) \;=\; 3.28419\,\sin\omega_1 t - 0.4050\,\sin\omega_3 t + 0.1808\,\sin\omega_5 t - \cdots
$$

Comparison of the above and Eq. (4.20) indicates that although the time function of Fig. 4-3(a) was represented by only 16 points, the agreement between Fourier transform and Fourier series is acceptable.

Example 4-2

As a second example, consider the function $F(t)$ of Fig. 4-3(b). In this case we again set $N = K = 16$ with $T = 16$. The results of calculation are shown in Table 4-2. The values of G given in this table should be

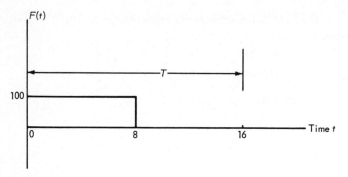

figure 4-3(b). *F(t) vs. time, Example 2.*

interpreted in conjunction with the inversion equation (4.6). Using the exponential form of G's

$$G_n = |G_n| e^{j\omega_n}$$

where $|G_n|$ is the absolute value of G_n, and ψ_n is the phase angle, Eq. (4.6) can be written

$$F(t_k) = \Delta\omega \sum_{n=0}^{N-1} |G_n| e^{j(\omega_n t_k + \psi_n)} \qquad (4.23)$$

The above equation represents the value of the function $F(t_k)$ for specific values of k. For example, for $k = 0$, $t_k = 0$, we have

$$F(0) = \Delta\omega \sum_{n=0}^{N-1} |G_n| e^{j\psi_n}$$

$$= \Delta\omega \sum_{n=0}^{N-1} (a_n + jb_n)$$

Substituting the values of a_n and b_n from Table 4-2, the above sum will result in the value of $F(0) = 100$.

The amplitude $\Delta\omega |G_n|$ of Eq. (4.23) can be plotted as a function of ω as shown in Fig. 4-4. Note that in this figure the foldover frequency

Table 4-2. Solution of Example Problem 4-2.

Time step k	Time t	Function $F(t)$
0	0	100
1	1	100
2	2	100
3	3	100
4	4	100
5	5	100
6	6	100
7	7	100
8	8	0
9	9	0
10	10	0
11	11	0
12	12	0
13	13	0
14	14	0
15	15	0

Freq. step k	Omega ω	Absolute value of G	Real part of G	Imaginary part of G
0	0	127.324	127.324	0
1	0.392699	81.5801	15.9155	−80.0126
2	0.785398	0	0	0
3	1.1781	28.6471	15.9155	−23.8192
4	1.5708	0	0	0
5	1.9635	19.1414	15.9155	−10.6344
6	2.35619	0	0	0
7	2.74889	16.2273	15.9155	− 3.16579
8	3.14159	0	0	0
9	3.53429	16.2273	15.9155	3.16579
10	3.92699	0	0	0
11	4.31969	19.1414	15.9155	10.6344
12	4.71239	0	0	0
13	5.10509	28.6471	15.9155	23.8192
14	5.49779	0	0	0
15	5.89049	81.5801	15.9155	80.0126

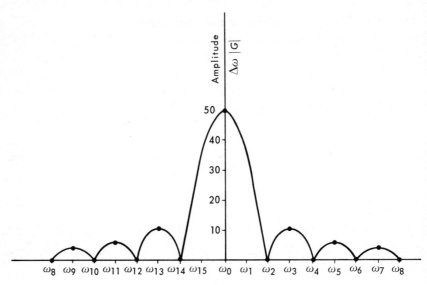

figure 4-4. *Amplitude-frequency plot of Fourier Transform of Example 2.*

occurs at ω_8 and the values of $\omega_9 = -\omega_7$, $\omega_{10} = -\omega_6$, ..., are plotted as negative frequencies.

COMPUTER PROGRAM FOR THE CALCULATION OF FOURIER TRANSFORM

The following computer program has been written in BASIC programming language [5] to implement the previously discussed calculations. The relation between the parameters of the program and those of the text are

Program (symbol)	Description	Text (symbol)
$K1$	Number of time steps	K
N	Number of frequency steps	N
T	Time span	T
$T1$	Time increment	Δt
$R(J)$	Real part of W^j	
$I(J)$	Imaginary part of W^j	
$F(J)$	Time function at $j\,\Delta t$	
$G(J)$	Real part of $G(\omega j)$	
$H(J)$	Imaginary part of $G(\omega j)$	
A	Absolute value of G	
$N1$	Foldover value of G	$N/2$

The loop 180-210 computes the required values of W. The loop 220-250 reads into the computer the time function $F(t)$ and incorporates the factor $T/2\pi K$ of Eq. (4.10). The loop 260-290 sets the values of G equal to F_0. The loop 300-640 is used to compute the modulus of nk by N (this can be done more efficiently by binary arithmetic, see footnote) and computes the required values of W multiplied by F in each column of matrix W. When the value of L becomes equal to $N/2$ (or $N1$ in the program) statement 380, the product $W^{N/2}F_i$ is set equal to $-W^0 F_i$, statement 420, and the complex conjugate values of previously obtained WF products are calculated, statements 480 and 490. When L becomes zero, statement 370, the values of W repeat and the instructions are transferred to statement 540. In the following loop 550-630 the calculated values of WF are incorporated into proper values of G. At the completion of statement 640, all the values of G's are calculated and the rest of the program simply prints these values.

```
FFFTTT           16:31         THURS. 04/11/68
100   DIM R(100),I(100),F(100),G(100),H(100),U(100),V(100)
110   READ K1,N,T
120   LET P1=3.1415927
130   LET P=(P1*2)/N
140   LET T1=T/K1
150   LET T2=P/T1
160   LET N1=N/2
170   LET T3=T1/(2*P1)
180   FOR J=0 TO N
```

The calculations of modulus of nk by N as given in statements 320-360 can be considerably simplified by using binary arithmetic. For example, for the row number $n = 7$ and the column number $k = 6$ the product nk will be 42, which in binary arithmetic is written

5	4	3	2	1	0	Power
32	16	8	4	2	1	Powers of 2
1	0	1	0	1	0	Product nk in binary bits (2 + 8 + 32)

$$\gamma = 3 \text{ bits}$$

The remainder is γ binary bits on the right of the product nk, where γ is obtained from $N = 2^\gamma$. In the above, the remainder is equal to 2.

```
190   LET R(J)=COS(P*J)
200   LET I(J)= -SIN(P*J)
210   NEXT J
220   FOR J=0 TO K1-1
230   READ F(J)
240   LET F(J)=F(J)*T3
250   NEXT J
260   FOR J=0 TO N-1
270   LET G(J)=F(0)
280   LET H(J)=0
290   NEXT J
300   FOR K=1 TO K1-1
310   FOR J=0 TO N-1
320   LET L=J*K
330   IF L=N THEN 540
340   IF L>N THEN 360
350   GO TO 380
360   LET L=L-INT(L/N)*N
370   IF L=0 THEN 540
380   IF L=N1 THEN 420
390   LET U(J)=R(L)*F(K)
400   LET V(J)=I(L)*F(K)
410   GO TO 510
420   LET U(J)= -U(0)
430   LET V(J)= -V(0)
440   LET J3=J
450   LET J3=J3-1
460   IF J3<1 THEN 510
470   LET J=J+1
480   LET U(J)=U(J3)
490   LET V(J)= -V(J3)
500   GO TO 450
510   NEXT J
520   LET J1=J
530   GO TO 550
540   LET J1=J-1
550   FOR J=0 TO N-1
560   FOR M=0 TO J1
570   LET G(J)=G(J)+U(M)
580   LET H(J)=H(J)+V(M)
590   LET J=J+1
600   IF J>N-1 THEN 640
610   NEXT M
620   LET J=J-1
630   NEXT J
```

```
640  NEXT K
650  PRINT "TIME STEP", "TIME", "FUNCTION"
660  FOR K=0 TO K1-1
670  PRINT K,K*T1,F(K)/T3
680  NEXT K
690  PRINT
700  PRINT "FREQ STEP","OMEGA","ABS VAL","REAL PART","IMAG PART"
710  PRINT
720  FOR J=0 TO N-1
730  LET A=SQR(G(J)↑2+H(J)↑2)
740  PRINT J,T2*J,A,G(J),H(J)
750  NEXT J
760  DATA 16,16,16
770  DATA 0,1,2,3,4,3,2,1,0,-1,-2,-3,-4,-3,-2,-1
780  END
```

COOLEY-TUKEY METHOD

The Cooley-Tukey method of computing the G matrix of Eq. (14.16) is based on expressing the W matrix of the same equation in terms of products of γ square matrices, where γ is related to the order of these matrices N by $N = 2^\gamma$. In this method the number of time and frequency steps are the same and are equal to N. Consider the expanded form of Eq. (4.11) for $N = 4$

$$
\begin{bmatrix} G_0 \\ G_1 \\ G_2 \\ G_3 \end{bmatrix} = \begin{bmatrix} W^{0,0} & W^{0,1} & W^{0,2} & W^{0,3} \\ W^{1,0} & W^{1,1} & W^{1,2} & W^{1,3} \\ W^{2,0} & W^{2,1} & W^{2,2} & W^{2,3} \\ W^{3,0} & W^{3,1} & W^{3,2} & W^{3,3} \end{bmatrix} \begin{bmatrix} F_0 \\ F_1 \\ F_2 \\ F_3 \end{bmatrix}
\tag{4.24}
$$

Taking the products of the exponents of W, the term nk, and obtaining their modulus for $N = 4$, we get

$$
\begin{bmatrix} G_0 \\ G_1 \\ G_2 \\ G_3 \end{bmatrix} = \begin{bmatrix} W^0 & W^0 & W^0 & W^0 \\ W^0 & W^1 & W^2 & W^3 \\ W^0 & W^2 & W^0 & W^2 \\ W^0 & W^3 & W^2 & W^1 \end{bmatrix} \begin{bmatrix} F_0 \\ F_1 \\ F_2 \\ F_3 \end{bmatrix}
\tag{4.25}
$$

Noting that the value of $W^0 = 1$, the W matrix of the above equation can be written as the product of two matrices ($\gamma = 2$) as follows:

$$
\begin{bmatrix} G_0 \\ G_2 \\ G_1 \\ G_3 \end{bmatrix} = \begin{bmatrix} 1 & W^0 & 0 & 0 \\ 1 & W^2 & 0 & 0 \\ 0 & 0 & 1 & W^1 \\ 0 & 0 & 1 & W^3 \end{bmatrix} \begin{bmatrix} 1 & 0 & W^0 & 0 \\ 0 & 1 & 0 & W^0 \\ 1 & 0 & W^2 & 0 \\ 0 & 1 & 0 & W^2 \end{bmatrix} \begin{bmatrix} F_0 \\ F_1 \\ F_2 \\ F_3 \end{bmatrix} \tag{4.26}
$$

The product of the above matrices will result in relation (4.25). Note that in Eq. (4.26) the order of the elements of the G matrix is changed. The product of the last two matrices of Eq. (4.26) is denoted as $[F]^1$ and is written

$$
[F]^1 = \begin{bmatrix} 1 & 0 & W^0 & 0 \\ 0 & 1 & 0 & W^0 \\ 1 & 0 & W^2 & 0 \\ 0 & 1 & 0 & W^2 \end{bmatrix} \begin{bmatrix} F_0 \\ F_1 \\ F_2 \\ F_3 \end{bmatrix} \tag{4.27}
$$

From the above, the elements of the $[F]^1$ matrix are

$$
\begin{aligned}
F_0{}^1 &= F_0 + W^0 F_2 \\
F_1{}^1 &= F_1 + W^0 F_3 \\
F_2{}^1 &= F_0 + W^2 F_2 \\
F_3 &= F_1 + W^2 F_3
\end{aligned} \tag{4.28}
$$

Having the elements of the $[F]^1$ matrix and using Eq. (4.26), the elements of the G matrix are computed.

The order of the elements of the G matrix in going from Eq. (4.25) to Eq. (4.26) is computed using binary arithmetic. The subscripts of these elements are first written in binary and then inverted to obtain the new order as follows:

Original Subscript	Binary form (γ binary bits)	Symmetrically inverted form	Final subscript
0	00	00	0
1	01	10	2
2	10	01	1
3	11	11	3

As a further example for $N = 8 = 2^3$, the above method will use 3 bit binary numbers resulting in the order of G's as given below:

Original Subscript	Binary form (γ binary bits)	Symmetrically inverted form	Final subscript
0	000	000	0
1	001	100	4
2	010	010	2
3	011	110	6
4	100	001	1
5	101	101	5
6	110	011	3
7	111	111	7

Multiplications and additions performed in obtaining the elements of $[F]^1$ matrix can be done by writing a computation tree and subsequently using binary arithmetic in computing the logic of operations.

For the four-by-four W matrix of Eq. (4.26) consider the tree of Fig. 4-5 where the original F matrix is placed on the left.

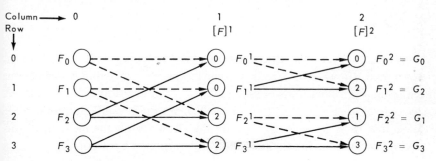

figure 4-5. *Computation tree of the Example.*

Solid lines of Fig. 4-5 represent multiplications, and dashed lines represent additions. The numbers in small circles are the powers of W. For example, $F_2{}^1$ is obtained by multiplying F_2 by W^2 and adding F_0 to it, as given by Eq. (4.28). The logic of determining the powers of W, which are entered in the circles, and the multiplications and additions are done using binary arithmetic as given below.

1. Assume N sampled values of $F(t)$ where $N = 2^\gamma$. The integer γ will equal the number of bits in the binary representation of N.

2. Arrange for γ columns of $[F]^1$, $[F]^2$, $[F]^3$, ..., $[F]^\gamma$ to be calculated.

3. Let the numbers in the circles representing the powers of W be denoted by p. For row k and column l (starting with k and l equal to zero), the procedure for computing p is as follows:

    a.   Write the binary number k.

    b.   Slide this number $\gamma - l$ places to the right filling the newly opened locations with zeros.

    c.   Reverse the order of the resulting binary bits. The resulting integer is p.

For example, for $N = 8 = 2^3 (\gamma = 3)$ the power in the circle of the 4th row ($k = 4$) and second column ($l = 2$) with $\gamma - l = 1$ is computed as follows:

    a.   The binary representation of k is 100.

    b.   Since $\gamma - l = 1$, shift k one bit to the right, resulting in the binary representation 010.

    c.   Reverse the order of (b), 010, which in integer form equals 2, that is, $p = 2$.

4. For multiplication and division denote the binary representation of row k by

$$k = k_{\gamma-1} \cdots k_2 k_1 k_0$$

where $k_0, k_1, \ldots$ are either zero or one. In column l, node k has a solid line drawn to it from a node in the $(l - 1)$ column having the same binary representation as k except the bit location $k_{\gamma-l}$ is replaced by one. For $N = 16 = 2^4$ ($\gamma = 4$) the element in the third column ($l = 3$), fifth row ($k = 5$), will have a solid line drawn from the element in the second column having the row number determined by writing

$$k = k_3 k_2 k_1 k_0 = 5 \quad \text{and in binary} = 0101$$

Replacing k_1 element by 1 we get row number = 0111 = 7, or row number seven. The dashed line comes from a node in the $(l - 1)$

column with binary representation of k, where $k_{\gamma - l}$ is replaced by zero. In the above example this will result in k of 5.

Using the above logic the computation tree of Fig. 4-6 for $N = 16$ can be constructed.

DISCUSSION OF FOURIER TRANSFORMS

Assuming that the function $F(t)$ is *real*, which is usually the case, the discrete values of its Fourier transform can be obtained by multiplying W and F matrices of Eq. (4.16). For N discrete values of Fourier transform G, this will take N^2 multiplications of complex numbers W's by real numbers F's. Taking advantage of the properties of the W matrix of Eq. (4.16) and the arrangement of product terms as given in Eq. (4.17), the number of multiplications can be considerably reduced. For example, for $N = 16$, the product matrices of Eq. (4.16) will require 256 multiplications of complex numbers by real numbers. Using the matrix calculations described by the six-step logic will reduce the number of multiplications to about 85. Note that this method does not require multiplication of complex numbers by complex numbers which account for four *real* multiplications.

The number of multiplications in the Cooley-Tukey method can be shown to be γ^N (where $N = 2^\gamma$). By this method for $N = 16$ ($\gamma = 4$) the number of multiplications will be 64. As can be seen from the computation tree of Cooley-Tukey method these multiplications involve complex numbers by complex numbers which should be counted as four real multiplications.

Since a general expression for the number of multiplications required in the matrix method is not available, a quantitative comparison of matrix and Cooley-Tukey method cannot be made. It is estimated, however, that the real computational advantage of Cooley-Tukey method is to be gained when N assumes very large values (in the order of $N = 2^8$ and above).

PROBLEMS

1. Show that G_1 and G_7 of Eq. (4.17) are complex conjugates.
2. Obtain the Fourier series of the function $F(t)$ of Fig. 4-3(a).
3. Obtain the real and imaginary parts of G_1 and G_2 of Table 4-1 by directly multiplying the proper elements of matrix equation (4.11).

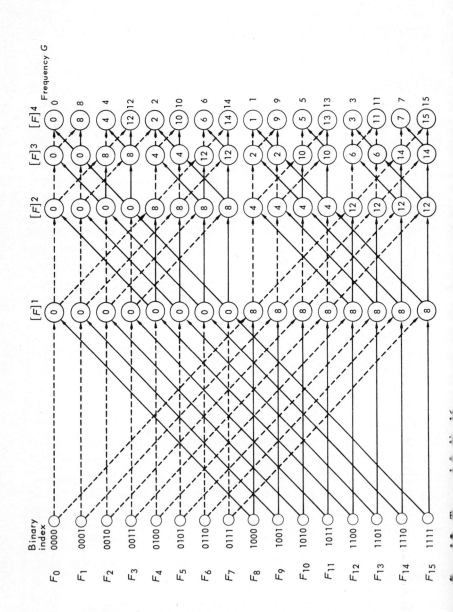

4. Setting $N = K = 8$, find the set of values $G_0, \ldots, G_7$ for $F(t)$ of Fig. 4-3(a).   *Answer:* $\Delta\omega$ = .392699, $|G_0|$ = $|G_2|$ = 0, $|G_1|$ = 4.34711, $|G_3|$ = .745846

5. Repeat the above problem for $N = K = 4$ and find the values $G_0, G_1, G_2,$ and G_3.   *Answer:* $\Delta\omega$ = .392699, $|G_0|$ = $|G_2|$ = 0, $|G_1|$ = 5.09296

6. Obtain the Fourier transform of

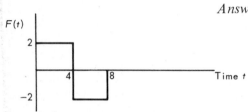

Answer: F_0 thru F_7 = 2, 2, 2, 2, –2, –2, –2, –2

$\Delta\omega$ = .785393

$|G_1|$ = 1.66357

$|G_3|$ = .689072

$|G_0|$ = $|G_2|$ = 0

by matrix method using $N = K = 8$.

7. Solve the above problem for $N = K = 4$.   *Answer:* $|G_1|$ = 1.80063

8. Obtain the Fourier transform of

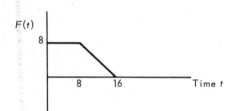

Answer: F_0 thru F_7 = 8, 8, 8, 8, 6, 4, 2, 0

$|G_0|$ = 14.0056

$|G_1|$ = 4.61834

$|G_4|$ = 1.27324

by matrix method using $N = K = 8$.

9. Obtain the Fourier transform of

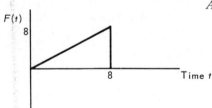

Answer: $N = K = 8$, $\Delta\omega$ = .785398

$|G_0|$ = 4.45634... $|G_4|$ = .63662

$N = K = 4$, $\Delta\omega$ = .785398

$|G_0|$ = 3.81972... $|G_2|$ = 1.27324

by matrix method using $N = K = 8$ and compare with $N = K = 4$.

10. Obtain the number of complex multiplications (complex numbers by real numbers) required for the computation of the G matrix of Eq. (4.17) for $N = K = 8$ and $N = K = 4$.

11. Obtain the Fourier transform of the function of Fig. 4-3(b) for $N = K = 8$.
12. Obtain Eq. (4.25) by performing the matrix multiplication of Eq. (4.26).
13. Verify that the values of $G_0, \ldots, G_3$ obtained from Fig. 4-5 are the same as those of Eq. (4.25).
14. Verify the additions, multiplications, and powers of W in the tree graph of Fig. 4-6.
15. Write a tree graph for $N = 8$.
16. Obtain the Fourier transform of Prob. 6 by the Cooley-Tukey method for $N = K = 4$.
17. Obtain the Fourier transform of Prob. 8 by the Cooley-Tukey method for $N = K = 8$.
18. Obtain the Fourier transform of Prob. 9 by the Cooley-Tukey method for $N = K = 8$.

REFERENCES

1. Brigham, E. O., and R. E. Morrow: The Fast Fourier Transform, *IEEE Spectrum,* December, 1967.
2. Danielson, G. C., and C. Lanczos: Some Improvements in Practical Fourier Analysis and Their Application to x-ray Scattering of Liquids, *J. Franklin Inst.,* vol. 233, pp. 365-380, 435-452, April, 1942.
3. Cooley, J. W., and J. W. Tukey: An Algorithm for the Machine Calculation of Complex Fourier Series, *Math. Computation,* vol. 19, pp. 297-301, April, 1965.
4. Cheng, David K.: "Analysis of Linear Systems," p. 133, Addison-Wesley Publishing Co., Inc., Reading, Mass., 1961.
5. BASIC Language, reference manual, General Electric Co., May, 1966.

5 NUMERICAL
METHODS AND CALCULUS
OF FINITE DIFFERENCES

In the numerical solution of mathematical problems, calculations are performed at discrete time or space intervals. The numerical values of the solutions are obtained at these intervals. This is to be distinguished from classical solutions where a continuous solution in the form of an equation is derived and it can be evaluated for the given values of the independent variable. To illustrate this, consider the second-order linear differential equation

$$\frac{d^2 y}{dt^2} + y = 0 \tag{5.1}$$

where t is time and the initial conditions are $y = 1$, $dy/dt = 0$ at

$t = 0$. The solution of this equation is of the form (see Appendix)

$$y = A \sin t + B \cos t$$

where A and B are constants, to be evaluated with initial conditions. This results in

$$y = \cos t \tag{5.2}$$

This solution gives the value of y for any value of time t. On the other hand, the numerical solution of differential equation (5.1) is obtained by selecting a time interval Δt transforming the equation into difference form (which will be discussed later) and obtaining a recursive equation for y in terms of two values of y at preceding time intervals. Or, in functional form

$$y_{i+1} = f(y_i, y_{i-1}, \Delta t) \tag{5.3}$$

the indexes i and $i-1$ refer to values of y at times t_i and t_{i-1} $(t_i - t_{i-1} = \Delta t)$ currently available and $i + 1$ refers to the value of y at t_{i+1} to be obtained. Thus, the application of the above relation together with the initial conditions will result in values of y at 0, Δt, $2\Delta t$, $3\Delta t$, ... and so on. It is important to note that, unlike solution (5.2), the value of y at $t = n\Delta t$ (where n is an integer) cannot be obtained unless the previous values of y are available.

It is seen that the numerical calculations involved in obtaining the value of y from Eq. (5.3) at a given time t are considerably more than the evaluation of solution (5.2). This is true in the above example since an analytical closed form solution exists for the differential equation (5.1). However, many engineering applications of numerical methods involve nonlinear differential equations, polynomial interpolation of tabulated functions, and the solution of transcendental equations where no analytical closed-form solutions exist. In these cases, solutions are obtained by numerical formulation of the problems and application of digital computers for evaluation of solutions.

It should be also pointed out that in most engineering problems, even if closed-form solutions are available, they should be evaluated and plotted while the numerical methods result in solution already in tabulated or graphical forms.

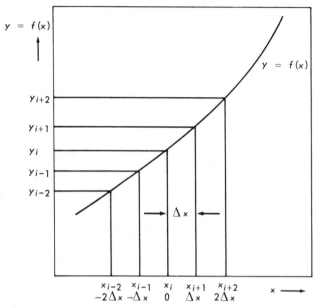

figure 5-1. *The function* $y = f(x)$.

DIFFERENCE OPERATORS

In numerical solutions, the difference operators are used in a manner similar to that of differential or Laplace operators (usually written as D or S) in the solution of differential equations. A complete familiarity with the difference operator is considered essential for an understanding of the underlying principles of numerical methods.

In order to define various difference operators, consider the function $y = f(x)$ described in Fig. 5-1 with equal intervals Δx for the independent variable x.

The first *backward difference* of y_i is defined as

$$\nabla y_i = y_i - y_{i-1} \tag{5.4}$$

where ∇ denotes the backward difference operator.

The second backward difference can be obtained from the first:

$$\nabla^2 y_i = \nabla(\nabla y_i) = \nabla(y_i - y_{i-1})$$

$$= \nabla y_i - \nabla y_{i-1} \tag{5.5}$$

$$= (y_i - y_{i-1}) - (y_{i-1} - y_{i-2})$$

$$= y_i - 2y_{i-1} + y_{i-2}$$

Similarly, the third backward difference becomes

$$\nabla^3 y_i = y_i - 3y_{i-1} + 3y_{i-2} - y_{i-3} \tag{5.6}$$

The first *forward difference* of y_i is defined as

$$\Delta y_i = y_{i+1} - y_i \tag{5.7}$$

where Δ denotes the forward difference operator. The second and third forward difference operators can be obtained from the first and second, respectively:

$$\Delta^2 y_i = \Delta(\Delta y_i)$$

$$= \Delta y_{i+1} - \Delta y_i$$

$$= y_{i+2} - 2y_{i+1} + y_i \tag{5.8}$$

$$\Delta^3 y_i = \Delta\left(\Delta^2 y_i\right)$$

$$= y_{i+3} - 3y_{i+2} + 3y_{i+1} - y_i$$

The first *central difference* operator of y_i is defined as

$$\delta y_i = y_{i+1/2} - y_{i-1/2} \tag{5.9}$$

where δ denotes the central difference operator. Note that the first central difference involves the values of the function at half intervals. The second central difference operator is obtained from the first:

$$\delta^2 y_i = \delta(\delta y_i) = \delta(y_{i+1/2} - y_{i-1/2})$$

$$= (y_{i+1/2+1/2} - y_{i+1/2-1/2}) - (y_{i-1/2+1/2} - y_{i-1/2-1/2})$$

$$= y_{i+1} - 2y_i + y_{i-1} \tag{5.10}$$

Similarly, third and fourth central difference operators can be obtained:

$$\delta^3 y_i = \delta\left(\delta^2 y_i\right) = y_{i+3/2} - 3y_{i+1/2} + 3y_{i-1/2} - y_{i-3/2}$$
$$\delta^4 y_i = \delta\left(\delta^3 y_i\right) = y_{i+2} - 4y_{i+1} + 6y_i - 4y_{i-1} + y_{i-2}$$

(5.11)

As a numerical example of the above equations, consider the calculation of difference operators of the equation $y = x^2$ as given in Table 5-1. In this table, the second and higher differences are obtained by arithmetical operations on the lower differences, for example, $\nabla^2 y = \nabla(\nabla y)$ or $\nabla^2 y = 5 - 3 = 2$. Only the value of $\delta^2 y$ is given since δy will require values of y at half interval points $(i \pm 1/2)$.

Table 5-1. Difference Operators of $y = x^2$

i	x_i	y_i	∇y_i	$\nabla^2 y_i$	$\nabla^3 y_i$	Δy_i	$\Delta^2 y_i$	$\Delta^3 y_i$	$\delta^2 y_i$
1	0					1	2	0	
2	1	1	1			3	2	0	2
3	2	4	3	2		5	2	0	2
4	3	9	5	2	0	7	2	0	2
5	4	16	7	2	0	9	2	0	2
6	5	25	9	2	0	11	2	0	2
7	6	36	11	2	0	13	2	0	2
8	7	49	13	2	0	15	2	0	2
9	8	64	15	2	0	17	2		2
10	9	81	17	2	0	19			2
11	10	100	19	2	0				

RELATIONSHIP BETWEEN DIFFERENCE AND DIFFERENTIAL OPERATORS

Let us define the symbolic difference operator E as follows:

$$Ey_i = y_{i+1}$$
$$E^{-1}y_i = y_{i-1}$$
$$E^{\pm 1/2}y_i = y_{i \pm 1/2}$$

(5.12)

Using these relations, we can obtain difference operators ∇, Δ, and δ in terms of E

$$\nabla y_i = y_i - y_{i-1} = y_i - E^{-1}y_i = \left(1 - \frac{1}{E}\right)y_i$$

$$\Delta y_i = y_{i+1} - y_i = Ey_i - y_i = (E - 1)y_i$$

$$\delta y_i = y_{i+1/2} - y_{i-1/2} = E^{1/2}y_i - E^{-1/2}y_i = (E^{1/2} - E^{-1/2})y_i$$

Equating both sides of these equations, we have

$$\nabla = 1 - \frac{1}{E} \qquad \text{backward difference operator}$$

$$\Delta = E - 1 \qquad \text{forward difference operator}$$

$$\delta = E^{1/2} - E^{-1/2} \qquad \text{central difference operator}$$

(5.13)

The following algebraic relations also hold between the operators:

$$\left.\begin{array}{l} \Delta k = k\,\Delta \\[4pt] Ek = kE \\[4pt] E\,\Delta = \Delta E \\[4pt] E^n\,\Delta = \Delta E^n \end{array}\right\} \quad \text{same for } \nabla \text{ and } \delta \text{ operators}$$

(5.14)

where k and n are constants.

The relation between difference operators discussed above and the differential operator is obtained by writing Taylor's series expansion of the function $f(x + h)$ about x. This results in [2]

$$f(x + h) = f(x) + hf'(x) + \frac{h^2}{2!}f''(x) + \cdots + \frac{h^n}{n!}f^n(x)$$

(5.15)

where primes denote derivatives with respect to x. Denoting the differential operator $D = d/dx$ and using the difference operator E, Eq. (5.15) becomes

$$Ef(x) = \left[1 + hD + \frac{h^2}{2!}D^2 + \frac{h^3}{3!}D^3 + \cdots\right]f(x)$$

(5.16)

Equating both sides of Eq. (5.16) and noting the infinite series in the brackets represents the series for the exponential function, we have

$$E = e^{hD}$$

or

$$D = \frac{\ln E}{h} \tag{5.17}$$

The above formulas can be applied to obtain relationships between difference and differential operators. These relationships are used in the numerical solution of differential equations, interpolation and numerical differentiation, and integration of tabular values.

RELATIONSHIP BETWEEN ∇ AND D

The backward difference operator ∇ can be obtained in terms of the differential operator D with the aid of Eq. (5.13):

$$\nabla = 1 - \frac{1}{E} = 1 - e^{-hD}$$

Taylor's series expansion of e^{-hD} will result.

$$\nabla = 1 - \left[1 - hD + \frac{h^2}{2!} D^2 - \frac{h^3}{3!} D^3 + \cdots \right]$$
$$= hD - \frac{h^2}{2!} D^2 + \frac{h^3}{3!} D^3 - \cdots \tag{5.18}$$

Similar expressions can be obtained for higher-order difference operators:

$$\nabla^2 = (1 - e^{-hD})^2 = 1 + e^{-2hD} - 2e^{-hD}$$

$$
\begin{aligned}
= 1 + \left(1 - \frac{2hD}{1!} + \frac{4h^2 D^2}{2!} - \cdots\right) \\
- 2\left(1 - \frac{hD}{1!} + \frac{h^2 D^2}{2!} - \frac{h^3 D^3}{3!} + \cdots\right)
\end{aligned}
\tag{5.19}
$$

$$= h^2 D^2 - h^3 D^3 + \frac{7}{12} h^4 D^4 - \cdots$$

The product of Eqs. (5.18) and (5.19) will give the value of ∇^3 in terms of differential operator D.

Conversely, the values of differential operator D can be obtained in terms of the difference operator ∇ by using the relation

$$e^{-hD} = 1 - \nabla$$
$$\ln e^{-hD} = -hD = \ln(1 - \nabla)$$
$$\tag{5.20}$$

Writing the series for $\ln(1 - \nabla)$, we have

$$D = \frac{1}{h}\left(\nabla + \frac{\nabla^2}{2} + \frac{\nabla^3}{3} + \frac{\nabla^4}{4} + \cdots\right) \tag{5.21}$$

Higher-order derivative operators $D^2, D^3, \ldots$, can be obtained successively by forming products of lower-order derivatives.

The relationships between ∇ and D operators are summarized below:

$$\nabla = hD - \frac{h^2}{2!} D^2 + \frac{h^3}{3!} D^3 - \cdots \tag{5.22a}$$

$$\nabla^2 = h^2 D^2 - h^3 D^3 + \frac{7}{12} h^4 D^4 - \cdots \tag{5.22b}$$

$$\nabla^3 = h^3 D^3 - \frac{3}{2} h^4 D^4 + \frac{5}{4} h^5 D^5 - \cdots \tag{5.22c}$$

$$D = \frac{1}{h}\left(\nabla + \frac{\nabla^2}{2} + \frac{\nabla^3}{3} + \frac{\nabla^4}{4} + \cdots\right) \tag{5.22d}$$

$$D^2 = \frac{1}{h^2}\left(\nabla^2 + \nabla^3 + \frac{11}{12}\nabla^4 + \frac{5}{6}\nabla^5 + \cdots\right) \qquad (5.22e)$$

$$D^3 = \frac{1}{h^3}\left(\nabla^3 + \frac{3}{2}\nabla^4 + \frac{7}{4}\nabla^5 + \cdots\right) \qquad (5.22f)$$

$$D^4 = \frac{1}{h^4}\left(\nabla^4 + 2\nabla^5 + \frac{17}{6}\nabla^6 + \cdots\right) \qquad (5.22g)$$

The above relations give the infinite series representations of backward differences in terms of differentials and vice versa. In using these equations in actual computation, however, we can only take a finite number of terms. Assuming that the series are convergent and that each term is smaller in magnitude than the preceding term (this is usually the case since the increment h is small and its powers increase), the first term dropped will represent the largest term neglected in the succeeding calculations. This term will give the "order of magnitude" of terms neglected in the finite difference representation. It should be noted that in truncating an infinite series the sum of all of the terms dropped gives the actual truncation error. Since getting an expression for this sum is a formidable task in many cases, the first term dropped from the series is taken as the "order of magnitude" of the error introduced. For example, taking the first two terms of Eq. (5.22b) and neglecting the terms $\frac{7}{12}h^4 D^4$ and beyond will result in an error equivalent to the absolute value of this term. In addition, as will be seen in the discussions of this and the following chapters, since in many problems the value of the derivative of the variable $D^4 y$ is also an unknown, this absolute value cannot be computed. For this reason we make another simplification in describing the error by stating that neglecting the third term and beyond of the series of Eq. (5.22b) introduces an "error of order h^4" in the resulting expression for ∇^2. The error of order h^4 is written symbolically as error of $O(h^4)$.

From Eq. (5.22) the derivatives of various functions can be obtained in terms of the available values of the function. Suppose that we want to obtain the value of D in terms of ∇'s with error of order h. From Eq. (5.22a) ∇^2 involves terms of h^2 and higher and since in Eq. (5.22d) this term, ∇^2, is divided by h, neglecting it in Eq. (5.22d) will result in an error of order h, as follows

$$D = \frac{\nabla}{h}, \quad Dy_i = \frac{1}{h}\nabla y_i$$

$$= \frac{y_i - y_{i-1}}{h} \qquad e = O(h)$$

(5.23)

where $e = O(h)$ refers to the error of order h. If the error is to be of order $h^2 [e = O(h^2)]$, the ∇^2 term in Eq. (5.22d) should be retained and the $\nabla^3, \nabla^4, \dots$ terms dropped. This, using definitions of ∇, ∇^2 results in

$$Dy_i = \frac{1}{h}\left(\nabla + \frac{\nabla^2}{2}\right)y_i$$

$$= \frac{1}{2h}(3y_i - 4y_{i-1} + y_{i-2}) \qquad e = O(h^2)$$

(5.24)

This result can also be obtained by setting $\nabla = 1 - E^{-1}$ from Eq. (5.13) and raising to appropriate powers, that is,

$$Dy_i = \frac{1}{h}\left(\nabla + \frac{\nabla^2}{2}\right)y_i$$

$$= \frac{1}{h}\left(1 - E^{-1} + \frac{(1 - E^{-1})^2}{2}\right)y_i$$

$$= \frac{1}{2h}(3y_i - 4y_{i-1} + y_{i-2})$$

The above relations can also be obtained for higher-order differentials. Table 5-2 summarizes the results.

RELATIONSHIP BETWEEN Δ AND D

The equations relating the forward difference operator Δ and the differential operator D are derived in a similar fashion as before. In this case from Eqs. (5.13) and (5.17), we have

Table 5-2. Relationship between Backward Difference Operator ∇ and Differential Operator D

	y_i	y_{i-1}	y_{i-2}	y_{i-3}	y_{i-4}	y_{i-5}	Error
$h\,Dy_i$	1	-1					
$h^2\,D^2y_i$	1	-2	1				
$h^3\,D^3y_i$	1	-3	3	-1			$e = O(h)$
$h^4\,D^4y_i$	1	-4	6	-4	1		
$2h\,Dy_i$	3	-4	1				
$h^2\,D^2y_i$	2	-5	4	-1			
$2h^3\,D^3y_i$	5	-18	24	-14	3		$e = O(h^2)$
$h^4\,D^4y_i$	3	-14	26	-24	11	-2	

$$E = 1 + \Delta = e^{hD}$$

$$D = \frac{1}{h}\ln(1 + \Delta) \tag{5.25}$$

resulting in

$$1 + \Delta = 1 + hD + \frac{h^2}{2!}D^2 + \frac{h^3}{3!}D^3 + \cdots$$

$$\Delta = hD + \frac{h^2}{2!}D^2 + \frac{h^3}{3!}D^3 - \cdots$$

and

$$D = \frac{1}{h}\left(\Delta - \frac{\Delta^2}{2} + \frac{\Delta^3}{3} - \frac{\Delta^4}{4} + \cdots\right) \tag{5.26}$$

Higher-order derivative and difference operators are obtained by taking various powers of Eq. (5.25). This results in the following second- and higher-order operator equations:

$$\Delta^2 = h^2 D^2 + h^3 D^3 + \frac{7}{12} h^4 D^4 + \cdots \tag{5.27a}$$

$$\Delta^3 = h^3 D^3 + \frac{3}{2} h^4 D^4 + \frac{5}{4} h^5 D^5 + \cdots \tag{5.27b}$$

$$D^2 = \frac{1}{h^2} \left(\Delta^2 - \Delta^3 + \frac{11}{12} \Delta^4 - \frac{5}{6} \Delta^5 + \cdots \right) \tag{5.27c}$$

$$D^3 = \frac{1}{h^3} \left(\Delta^3 - \frac{3}{2} \Delta^4 + \frac{7}{4} \Delta^5 - \cdots \right) \tag{5.27d}$$

$$D^4 = \frac{1}{h^4} \left(\Delta^4 - 2\Delta^5 + \frac{17}{6} \Delta^6 - \cdots \right) \tag{5.27e}$$

Equations (5.26) and (5.27) can be used to find the differentials of functions in terms of available functional values. For example, neglecting errors of order h and higher, $D^2 y_i$ can be obtained from Eq. (5.27c). Dropping Δ^3 and higher-order terms from this equation and using Eq. (5.8), we have

$$
\begin{aligned}
D^2 y_i &= \frac{1}{h^2} \Delta^2 y_i \\
&= \frac{1}{h^2} (y_{i+2} - 2y_{i+1} + y_i)
\end{aligned}
\tag{5.28}
$$

Using this procedure a table of coefficients, corresponding to Table 5-2, can be calculated for forward-difference operators. This is given in Table 5-3. Note that the second relation above can be derived from the first by substituting $\Delta = 1 - E$ from Eq. (5.13) and using the definition of difference operator, that is,

$$
\begin{aligned}
D^2 y_i &= \frac{1}{h^2} (E - 1)^2 y_i \\
&= \frac{1}{h^2} (1 - 2E + E^2) y_i \\
&= \frac{1}{h^2} (y_i - 2y_{i+1} + y_{i+2})
\end{aligned}
$$

Table 5-3. Relationship between Forward Difference Operator Δ and Differential Operator D

	y_i	y_{i+1}	y_{i+2}	y_{i+3}	y_{i+4}	y_{i+5}	Error
$h\,Dy_i$	-1	1					
$h^2\,D^2 y_i$	1	-2	1				
$h^3\,D^3 y_i$	-1	3	-3	1			$e = O(h)$
$h^4\,D^4 y_i$	1	-4	6	-4	1		
$2h\,Dy_i$	-3	4	-1				
$h^2\,D^2 y_i$	2	-5	4	-1			
$2h^3\,D^3 y_i$	-5	18	-24	14	-3		$e = O(h^2)$
$h^4\,D^4 y_i$	3	-14	26	-24	11	-2	

RELATIONSHIP BETWEEN δ AND D OPERATORS

The equations relating the central difference operator δ and differential operator D can be obtained using Eqs. (5.13) and (5.17).

$$\delta = E^{1/2} - E^{-1/2}, \quad E = e^{hD}, \quad D = \frac{1}{h}\ln E \qquad (5.29)$$

In addition, let us define the mean operator μ as follows:

$$\mu = \frac{1}{2}(E^{1/2} + E^{-1/2}) \qquad (5.30)$$

This will result in

$$\mu\delta = \frac{1}{2}(E - E^{-1}) = \frac{1}{2}(e^{hD} - e^{-hD}) = \sinh(hD) \qquad (5.31)$$

$$\mu\delta(y_i) = \frac{1}{2}(y_{i+1} - y_{i-1})$$

and

$$\mu^2 = \frac{1}{4}(E + E^{-1} + 2) = 1 + \frac{\delta^2}{4} \qquad (5.32)$$

Substituting the Taylor series expansion for $\sinh(hD)$, Eq. (5.31) becomes

$$\mu\delta = hD + \frac{h^3 D^3}{6} + \frac{h^5 D^5}{120} + \cdots \tag{5.33}$$

Using the expression for δ from Eq. (5.29), squaring and substituting $E = e^{hD}$, we have

$$\delta^2 = (E - 2 + E^{-1}) = 2\left(\frac{e^{hD} + e^{-hD}}{2} - 1\right) \tag{5.34}$$

$$= 2[\cosh(hD) - 1]$$

Substituting the series for $\cosh(hD)$ results in

$$\delta^2 = h^2 D^2 + \frac{h^4 D^4}{12} + \frac{h^6 D^6}{360} + \cdots \tag{5.35}$$

The general expression for δ^n, where $n = 1, 2, 3, \ldots$, can be obtained from Eq. (5.29)

$$\delta = e^{hD/2} - e^{-hD/2} = 2\sinh\frac{hD}{2}$$

$$\delta^n = 2^n \sinh^n \frac{hD}{2} \tag{5.36}$$

Using combinations of the above relations, we can obtain expressions for higher powers of central difference operators. For example, combination of Eqs. (5.33) and (5.35) gives

$$\mu\delta^3 = (\mu\delta)(\delta^2) = \left(hD + \frac{h^3 D^3}{6} + \frac{h^5 D^5}{120} + \cdots\right)$$

$$\left(h^2 D^2 + \frac{h^4 D^4}{12} + \frac{h^6 D^6}{360} + \cdots\right) \tag{5.37}$$

$$= h^3 D^3 + \frac{h^5 D^5}{4} + \frac{h^7 D^7}{40} + \cdots$$

Squaring the first few terms of Eq. (5.35) will result in

$$\delta^4 = h^4 D^4 + \frac{h^6 D^6}{6} + \frac{h^8 D^8}{80} + \cdots \tag{5.38}$$

The differential operators D can also be obtained in terms of central difference operators δ.

For example, from Eqs. (5.31) and (5.32) we get

$$D = \frac{1}{h} \sinh^{-1} \mu \delta = \frac{1}{h} \left(\mu \delta - \frac{\mu \delta^3}{6} + \frac{\mu \delta^5}{30} - \cdots \right) \tag{5.39}$$

where the right side of the equation is obtained from Taylor series expansion of $\sinh^{-1} \mu \delta$. Similarly,

$$D^2 = \frac{1}{h^2} \left(\delta^2 - \frac{\delta^4}{12} + \frac{\delta^6}{90} - \cdots \right) \tag{5.40a}$$

$$D^3 = \frac{\mu}{h^3} \left(\delta^3 - \frac{\delta^5}{4} + \frac{7\delta^7}{120} - \cdots \right) \tag{5.40b}$$

$$D^4 = \frac{1}{h^4} \left(\delta^4 - \frac{\delta^6}{6} + \frac{7\delta^8}{240} - \cdots \right) \tag{5.40c}$$

Note that in the above relations, only odd differentials (D and D^3) contain terms involving μ's.

Using the above relations, a table can be obtained relating differential and functional values. For example, the second differential D^2, neglecting $e = O(h^2)$, can be obtained from Eq. (5.40a). From Eq. (5.38), δ^4 involves terms of h^4 and higher; since this term (δ^4) in Eq. (5.40a) is divided by h^2, neglecting this will result in errors of order h^2 and higher. Using Eq. (5.10), this results

$$D^2 y_i = \frac{1}{h^2} (\delta^2) y_i = \frac{1}{h^2} (y_{i+1} - 2y_i + y_{i-1}) \qquad e = O(h^2) \tag{5.41}$$

This result can also be obtained by substituting the value of $\delta = E^{1/2} - E^{-1/2}$ from Eq. (5.13) and using the definition of operator E, that is,

$$D^2 y_i = \frac{1}{h^2} (\delta^2) y_i$$

$$= \frac{1}{h^2} (E^{1/2} - E^{-1/2})^2 y_i$$

$$= \frac{1}{h^2} (E - 2 + E^{-1}) y_i$$

$$= \frac{1}{h^2} (y_{i+1} - 2y_i + y_{i-1})$$

Table 5-4 gives the relation between various orders of differential operators and central difference operators.

The forward and backward differences are primarily used in incorporating the initial and/or boundary conditions (as will be seen in Chaps. 6 and 7) in the finite difference representation of the differential equation while the central differences are used for getting the general finite difference formulation.

Example. As an example of application of the above equations, consider the numerical solution of the differential equation

Table 5-4. Relationship between Central Difference Operator δ and Differential Operator D

	y_{i-3}	y_{i-2}	y_{i-1}	y_i	y_{i+1}	y_{i+2}	y_{i+3}	
$2h\,Dy_i$			-1	0	1			
$h^2 D^2 y_i$			1	-2	1			$e = O(h^2)$
$2h^3 D^3 y_i$		-1	2	0	-2	1		
$h^4 D^4 y_i$		1	-4	6	-4	1		
$12h\,Dy_i$		1	-8	0	8	-1		
$12h^2 D^2 y_i$		-1	16	-30	16	-1		$e = O(h^4)$
$8h^3 D^3 y_i$	1	-8	13	0	-13	8	-1	
$6h^4 D^4 y_i$	-1	12	-39	56	-39	12	-1	

$$\frac{d^2y}{dt^2} + \frac{dy}{dt} + y = 0 \tag{5.42}$$

ubject to initial conditions, $y = 1$, $dy/dt = 0$ at $t = 0$. This equation in perator form can be written

$$D^2 y_i + D y_i + y_i = 0 \tag{5.43}$$

·here $D = d/dt$ and the initial conditions are $y_0 = 1$, $Dy_0 = 0$. Using he central differences of order h^2, from Table 5-4, we have

$$D^2 y_i = \frac{y_{i-1} - 2y_i + y_{i+1}}{h^2} \qquad e = O(h^2) \tag{5.44}$$

$$Dy_i = \frac{y_{i+1} - y_{i-1}}{2h} \qquad e = O(h^2)$$

·here $h = \Delta t$ is the assumed time interval. Substituting Eq. (5.44) in :q. (5.43), we obtain a recursion relation for y_{i+1}, as follows:

$$y_{i+1} = \frac{1}{2+h} \left[(4 - 2h^2) y_i + (h - 2) y_{i-1} \right] \tag{5.45}$$

·his equation gives the value of y_{i+1} provided that the two previous alues of y are available, that is, y_i and y_{i-1}. To start the problem, wo values of y will be needed. These will be obtained from initial ·onditions. Using central differences for the initial condition $Dy_0 = 0$, ·e have

$$Dy_0 = \frac{y_1 - y_{-1}}{2h} = 0 \qquad y_{-1} = y_1 \tag{5.46}$$

·Jote that y_{-1} in the above relation is inserted to make the derivative at ·0 symmetrical. It actually does not exist in calculations. In addition, ·e have the initial condition given in the problem $y_0 = 1$. Using this condition together with Eq. (5.46) and setting $i = 0$ in Eq. (5.45), w can solve for y_1. Having the values of y_0 and y_1, Eq. (5.45) will give th value of $y_2 (i = 1)$. Thus, the solution can be continued for $i = 3,4, \ldots$ resulting in values of y at discrete time steps $n \Delta t$ where $n = 0,1,2,3, \ldots$

Mathematically, one of the important problems in the numerica solution of differential equations is choosing the value $h = \Delta t$. If Δt i too small (in addition to the problems of the stability of solution which may arise), a great number of iterations will be required to obtain th value of y for a given time. For example, if the value of y at $t = 10$ i required, a $\Delta t = 0.01$ will require 1,000 applications of Eq. (5.45), whil a $\Delta t = 0.1$ will require only 100 applications of this equation for obtainin the desired value of y. On the other hand, if Δt is too large since th error is of the order of Δt^2 (or h^2 in Eq. (5.44)), the error in eacl application of Eq. (5.45) may accumulate, resulting in large errors in th solution. Some theoretical background will be given for calculating th effect of Δt on the size of accumulated error in Chap. 7. However, i most engineering applications, the value of Δt is chosen by numerica experiments. This means that a value of Δt is chosen and the problen run on the computer and solutions obtained. The value of Δt is ther varied and the solutions examined. In this way, a value is chosen tha is a tradeoff between computation time and the desired accuracy of th solution.

DERIVATIVE OF A TABULATED FUNCTION

The derivatives of tabulated functions can be obtained with the aid o relations already derived. These derivatives are represented in terms of backward, forward, and central differences in Tables 5-2, 5-3, and 5-4 respectively. For example, the central difference approximation o order h^2 for the second derivative of a function is given in Table 5-4 a

$$h^2 D^2 y_i = y_{i-1} - 2y_i + y_{i+1}$$

Applying this formula to tabulated values of $y = \log x$ of Table 5-5 of page 210, we can obtain derivatives at various values of x. For $x = 1.02$, we have

$$D^2 \log x \big|_{x = 1.02} = \frac{0.0043214 - 2(0.0086002) + 0.0128372}{(0.01)(0.01)} \qquad (5.47)$$

$$= -4.18 \times 10^{-9}$$

INTEGRAL OF A TABULATED FUNCTION

Using the symbol for the differential operator D, we can define an integral operator as the inverse of D:

$$D^{-1}f(x) = \text{Integral of } f(x) \tag{5.48}$$

From this,

$$\int_a^{a+h} f(x)\, dx = \left.\frac{f(x)}{D}\right|_a^{a+h} = \frac{1}{D}[f(a+h) - f(a)]$$

$$= \frac{E-1}{D}f(a)$$

Similarly, for the integration interval between a and $a + nh$, we have

$$\int_a^{a+nh} f(x)\, dx = \left.\frac{f(x)}{D}\right|_a^{a+nh} = \frac{E^n - 1}{D}f(a) \tag{5.49}$$

Since the integration interval in the above equation starts at a and increases to $a + nh$, it is reasonable to derive the integral in terms of forward differences. Using the relations

$$E = 1 + \Delta$$

$$D = \frac{1}{h}\left(\Delta - \frac{\Delta^2}{2} + \frac{\Delta^3}{3} - \cdots\right)$$

in Eq. (5.49) and dividing the numerator by the denominator, we have

$$\int_a^{a+nb} f(x)\, dx$$

$$= nh\left[1 + \frac{n}{2}\Delta + \frac{n(2n-3)}{12}\Delta^2 + \frac{n(n-2)^2}{24}\Delta^3 + \cdots\right]f(a) \tag{5.50}$$

For $n = 1$, neglecting terms involving Δ^2 and higher, we have

$$\int_a^{a+b} f(x)\, dx = h\left[1 + \frac{\Delta}{2}\right] f(a)$$

$$= h\left[f(a) + \frac{f(a + h) - f(a)}{2}\right] \qquad (5.51)$$

$$= \frac{h}{2}[f(a) + f(a + h)]$$

which can be recognized as the *trapezoidal* rule of integration. For $n = 2$, the term involving Δ^3 in Eq. (5.50) drops out. Neglecting terms involving Δ^4 and higher, we have

$$\int_a^{a+2b} f(x)\, dx = h\left(2 + 2\Delta + \frac{\Delta^2}{3}\right) f(a) \qquad (5.52)$$

This can be readily evaluated by using the relation $\Delta = E - 1$. This results in

$$\int_a^{a+2b} f(x)\, dx = h\left(2 + 2(E - 1) + \frac{(E - 1)^2}{3}\right) f(a)$$

$$= \frac{h}{3}(1 + 4E + E^2) f(a) \qquad (5.53)$$

$$= \frac{h}{3}[f(a) + 4f(a + h) + f(a + 2h)]$$

which can be recognized as *Simpson's one-third* rule of integration. Similarly, setting $n = 3$ in Eq. (5.50) and transforming in E operator form,

$$\int_a^{a+3b} f(x)\, dx = \tfrac{3}{8} h[f(a) + 3f(a + h) + 3f(a + 2h) + f(a + 3h)]$$

$$(5.54)$$

This equation is usually referred to as *Simpson's three-eighths* rule.

Integrals of tabulated functions can also be obtained in terms of backward and central differences. The central difference relations can be evaluated by writing the integral of the function $f(x)$ in a symmetrical region

$$\int_{a-nb}^{a+nb} f(x)\, dx = \frac{f(x)}{D}\Bigg|_{a-nh}^{a+nh} = \frac{f(a+nh) - f(a-nh)}{D}$$

$$= \frac{E^n - E^{-n}}{D} f(a)$$

(5.55)

For $n = 1$, this relation, from Eqs. (5.29), (5.30), and (5.39), can be written

$$\int_{a-b}^{a+b} f(x)\, dx = \frac{(E^{1/2} - E^{-1/2})(E^{1/2} + E^{-1/2})}{D} f(a)$$

$$= \frac{2\delta\mu}{D} f(a)$$

$$= \frac{2h\,\delta\mu}{\delta\mu - \dfrac{\delta^3 \mu}{6} + \dfrac{\delta^5 \mu}{30} - \cdots} f(a)$$

(5.56)

which, upon dividing, results in

$$\int_{a-b}^{a+b} f(x)\, dx = 2h\left(1 + \frac{\delta^2}{6} + \frac{2}{15}\delta^4 + \cdots\right) f(a)$$

(5.57)

Neglecting the terms of δ^4 and higher gives

$$\int_{a-b}^{a+b} f(x)\, dx = 2h\left(1 + \frac{\delta^2}{6}\right) f(a)$$

$$= \frac{h}{3}\left[f(a-h) + 4f(a) + f(a+h)\right]$$

(5.58)

which is Simpson's one-third rule of integration in central difference form.

For $n = 2$, Eq. (5.55) results in

$$\int_{a-2h}^{a+2h} f(x)\, dx = \frac{(E^1 - E^{-1})(E^1 + E^{-1})}{D} f(a)$$

$$= \frac{(2h\, \delta\mu)(\delta^2 + 2)^*}{\delta\mu - \dfrac{\delta^3 \mu}{6} + \dfrac{\delta^5 \mu}{30} - \cdots} \qquad (5.59)$$

$$= 2h\left(2 + \frac{4}{3}\delta^2 + \frac{7}{45}\delta^4 + \cdots\right)$$

Neglecting the terms of fourth and higher orders in δ, we get

$$\int_{a-2h}^{a+2h} f(x)\, dx = \frac{2h}{3}\left[4f(a - h) - 2f(a) + 4f(a + h)\right] \qquad (5.60)$$

Note that all integral formulas should give correct answers for $f(x) = C$ constant. This is usually used as a check of the integral formulas. For example, Eq. (5.60) for $f(x) = C$ results in

$$\int_{a-2h}^{a+2h} C\, dx = 4hC = \frac{2h}{3}(4C - 2C + 4C) = 4hC \qquad (5.61)$$

Integrals of functions can similarly be obtained in terms of backward differences. For this, consider integration over the interval $a - nh$ to a

$$\int_{a-nh}^{a} f(x)\, dx = \frac{f(x)}{D}\Bigg|_{a-nh}^{a} = \frac{f(a) - f(a - nh)}{D}$$

$$= \frac{1 - E^{-n}}{D} f(a)$$

$$= \frac{1 - (1 - \nabla)^n}{D} f(a)$$

(continued)

---

*This expression is obtained by writing $\delta = E^{1/2} - E^{-1/2}$ and squaring $\delta^2 = E + E^{-1} - 2$. This gives $E + E^{-1} = \delta^2 + 2$.

$$= \frac{n\nabla - \dfrac{n(n-1)}{2!}\nabla^2 + \dfrac{n(n-1)(n-2)}{3!}\nabla^3 - \cdots}{\dfrac{1}{h}\left(\nabla + \dfrac{\nabla^2}{2} + \dfrac{\nabla^3}{3} + \cdots\right)} f(a)$$

$$(5.62)$$

Division of this expression will result in

$$\int_{a-nb}^{a} f(x)\, dx = h\left(n - \frac{n^2}{2}\nabla + \frac{n^3}{6}\nabla^2 - \cdots\right) f(a)$$

This expression can be evaluated for different values of n.

INTEGRATION FORMULAS BY INTERPOLATING POLYNOMIALS

The integration formulas given above and additional integration formulas for the numerical solution of differential equations can be obtained by using polynomial interpolations. The basic idea is that a polynomial with unknown coefficients is assumed to pass through the sets of given points. Using the values of these points, the coefficients of the polynomial can be evaluated. These coefficients are then used to obtain integration formulas.

Consider Fig. 5-1 of page 185 with variables x and $y = f(x)$. Assume that the straight line

$$y = Ax + B \qquad (5.63)$$

passes through (x_i, y_i) and (x_{i+1}, y_{i+1}). For simplicity, let us set $x_i = 0$ or equivalently take x_i as the starting point and set the points $(0, y_i)$ and $(\Delta x, y_{i+1})$ in Eq. (5.63). This results in

$$y_i = B$$
$$y_{i+1} = A(\Delta x) + B \qquad (5.64)$$

Solving these equations for A and B, we obtain

$$A = \frac{y_{i+1} - y_i}{\Delta x}$$

(5.65)

$$B = y_i$$

Using these values, Eq. (5.63) becomes

$$y = \frac{y_{i+1} - y_i}{\Delta x} x + y_i$$

(5.66)

The area under the curve from $x = 0$ to $x = \Delta x$ is obtained by integration of Eq. (5.66). This gives

$$A = \int_0^{\Delta x} y \, dx = \frac{y_{i+1} - y_i}{\Delta x} \left(\frac{\Delta x^2}{2}\right) + y_i \Delta x = \frac{\Delta x}{2} (y_{i+1} + y_i)$$

(5.67)

which is the *trapezoidal law* of integration. In a similar fashion, the Simpson rule of integration can be obtained by assuming the second degree polynomial

$$y = Ax^2 + Bx + C$$

(5.68)

passing through the points $(-\Delta x, y_{i-1})$, $(0, y_i)$, $(\Delta x, y_{i+1})$. Substituting these values in Eq. (5.68) and solving for A, B, and C, we obtain

$$A = \frac{y_{i+1} - 2y_i + y_{i-1}}{2(\Delta x)^2}, \quad B = \frac{y_{i+1} - y_{i-1}}{2\Delta x}, \quad C = y_i$$

(5.69)

Using these values of A, B, and C in Eq. (5.68) and integrating,

$$\text{Area} = \int_{-\Delta x}^{\Delta x} y \, dx = \frac{\Delta x}{3} (y_{i+1} + 4y_i + y_{i-1})$$

(5.70)

which is *Simpson's one-third* rule of integration.

Expressions (5.67) and (5.70) give areas under the curve between points which satisfy expressions (5.63) and (5.68). It is sometimes desirable, as will be seen in predictor-corrector methods of solving differential equations (Chap. 6), to extend the range of integration using the same polynomials. For example, using Eq. (5.68) with the coefficients (5.69), the area under the curve of Fig. 5-1 from $-2\Delta x$ to $2\Delta x$ becomes

$$\int_{-2\Delta x}^{2\Delta x} y \, dx = \frac{4\Delta x}{3} (2y_{i+1} - y_i + 2y_{i-1}) \tag{5.71}$$

This expression gives the area under the curve extending from (x_{i-2}, y_{i-2}) to (x_{i+2}, y_{i+2}) in terms of values of y extending from y_{i-1} to y_{i+1}.

Differentiating Eq. (5.68) twice will result in

$$\frac{d^2 y}{dx^2} = 2A \tag{5.72}$$

Substituting the value of A from Eq. (5.69), we obtain

$$\frac{d^2 y}{dx^2} = \frac{y_{i+1} - 2y_i + y_{i-1}}{\Delta x^2} \tag{5.73}$$

which is the central difference expression for the second derivative of y at x_i.

Differentiating Eq. (5.68) once, we obtain

$$\frac{dy}{dx} = 2Ax + B \tag{5.74}$$

The value of this derivative at $x = x_i = 0$ is obtained by substituting B from Eq. (5.69) and $x = 0$. This results in

$$\frac{dy}{dx} = \frac{y_{i+1} - y_{i-1}}{2\Delta x} \tag{5.75}$$

which is the *central difference* expression for the first derivative.

The integrals derived here, in addition to being used in obtaining the integrals of tabulated functions, are used in the numerical solution of differential equations, as will be discussed in Chap. 6.

GREGORY-NEWTON INTERPOLATION FORMULA

Consider the function $f(x)$ tabulated from $x = a$ to $x = b$ at incre ments of $\Delta x = h$. It is desired to find the value of $f(x)$ at $x = a + nh$ where n can be a fraction. Using the relation

$$f(a + nh) = E^n f(a) \tag{5.76}$$

and the expression for E in terms of forward differences, we have

$$E^n f(a) = f(a + nh) = (1 + \Delta)^n f(a)$$

$$= \left[1 + n\,\Delta + \frac{n(n-1)}{2!} \Delta^2 + \frac{n(n-1)(n-2)}{3!} \Delta^3 + \cdots \right] f(a) \tag{5.77}$$

The above equation can also be derived in terms of backward differences ∇. For this we start from point b and write

$$E^{-n} f(b) = f(b - nh)$$

$$= (1 - \nabla)^n f(b) \tag{5.78}$$

$$= \left[1 - n\,\nabla + \frac{n(n-1)}{2!} \nabla^2 - \frac{n(n-1)(n-2)}{3!} \nabla^3 + \cdots \right] f(b)$$

Equations (5.77) and (5.78) are referred to as Gregory-Newton inter polation formulas. Using these relations the values of tabulated functions can be obtained at any point within the interval by obtaining the required forward and backward differences.

Example. Consider the table of $\log_{10} x$ for values of x between 1.0 and 1.05 at increments of $\Delta x = h = 0.01$ to be given. It is desired to obtain $\log x$ at $x = 1.024$ by forward interpolation formula (5.77).

The necessary calculations are represented in Table 5-5. Since the value of $\log x$ at 1.024 is desired, let us set $f(a) = 1.02$. The value of n is

$$n = \frac{1.024 - 1.020}{0.010} = 0.400$$

Using this value of n together with proper values of Table 5-5 in Eq. (5.77), and neglecting terms of fourth order and higher, we have

$$f(1.02 + 0.4 \times 0.01) = 0.0086002 + 0.4(0.0042370)$$

$$+ \frac{0.4(0.4 - 1)}{2}(-0.0000409) \qquad (5.79)$$

$$+ \frac{0.4(0.4 - 1)(0.4 - 2)}{6}(0.0000008)$$

From this

$$f(1.024) = 0.0103000$$

Note that the first two terms on the right of the equal sign in Eq. (5.79) represent the familiar linear interpolation terms.

STERLING'S INTERPOLATION FORMULA

Just as Gregory-Newton interpolation formulas are based on forward and backward differences, the Sterling interpolation formula is based on central differences. For the derivation of this, consider the Taylor series expansion of the function $f(a + nh)$

$$f(a + nh) = \left(1 + nhD + \frac{n^2}{2!}h^2D^2 + \frac{n^3}{3!}h^3D^3 + \cdots\right)f(a) \qquad (5.80)$$

where, as before, $f(a + nh)$ is the value of tabulated function at $a + nh$, h is the increment of independent variable x, and n can be a fraction. Substituting the values of D in terms of central difference operator δ, from Eqs. (5.39) and (5.40) in Eq. (5.80), we have

$$f(a + nh) = \left[1 + n\mu\left(\delta - \frac{\delta^3}{6} + \frac{\delta^5}{30} - \cdots\right) + \frac{n^2}{2!}\left(\delta^2 - \frac{\delta^4}{12} + \cdots\right)\right.$$

$$\left. + \frac{n^3}{3!}\mu\left(\delta^3 - \frac{\delta^5}{4} + \cdots\right) + \cdots\right]f(a) \qquad (5.81)$$

Combining powers of δ results in

Table 5-5. Example of Gregory-Newton Interpolation Formula

x	$f(s) = \log x$	$\Delta f(x)$	$\Delta^2 f(x)$	$\Delta^3 f(x)$
1.00	0.0000000	0.0043214	−0.0000426	0.0000008
1.01	0.0043214	0.0042788	−0.0000418	0.0000009
1.02	0.0086002	0.0042370	−0.0000409	0.0000008
1.03	0.0128372	0.0041961	−0.0000401	
1.04	0.0170333	0.0041560		
1.05	0.0211893			

Table 5-6. Example of Sterling's Interpolation Formula

x	$f(x) = \log x$	$\mu\,\delta f(x)$	$\delta^2 f(x)$	$\mu\,\delta^3 f(x)$
1.00	0.0000000			
1.01	0.0043214	0.0043001	−0.0000426	
1.02	0.0086002	0.0042579	−0.0000418	0.0000008
1.03	0.0128372	0.0042165	−0.0000409	0.0000008
1.04	0.0170333	0.0041760	−0.0000401	
1.05	0.0211893			

$$f(a + nh) = \left[1 + n\mu\,\delta + \frac{n^2}{2!}\delta^2 + \frac{n(n^2 - 1)}{3!}\mu\,\delta^3 + \frac{n^2(n^2 - 1)}{4!}\delta^4 \right.$$
$$\left. + \frac{n(n^2 - 1)(n^2 - 4)}{5!}\mu\,\delta^5 + \cdots \right] f(a) \tag{5.82}$$

Equation (5.82) is the *Sterling interpolation* formula. This equation can be used for finding the values of the function at any point within the intervals by obtaining the required differences.

Example. As an example of application of this formula, let us again consider the problem of evaluating log 1.024 from tabulated values of $\log x$ as given in Table 5-5. The necessary values to be used in Eq. (5.82) are given in Table 5-6. The values of $\mu\,\delta$, δ^2, and $\mu\,\delta^3$ are obtained from the following (previously given) equations:

$$\mu\,\delta y_i = \frac{1}{2}(y_{i+1} - y_{i-1})$$
$$\delta^2 y_i = y_{i+1} - 2y_i + y_{i-1} \tag{5.83}$$
$$\mu\,\delta^3 y_i = (\mu\,\delta)(\delta^2 y_i)$$

The value of n as obtained previously is 0.4. Using this value of n and the tabulated values of Table 5-6 and neglecting terms of fourth degree and higher in Eq. (5.82) results in

$$f(1.024) = 0.0086002 + 0.4(0.0042579) + \frac{(0.4)^2}{2}(-0.0000418)$$

$$+ \frac{0.4(0.16 - 1)}{6}(0.0000008) = 0.0103000$$

(5.84)

where $f(a) = 0.0086002$ is used as the starting point. This result agrees to seven decimal places with the result from the previous example.

PROBLEMS

1. Derive the expression for $\Delta^3 y_i$ as given by Eq. (5.8).
2. Derive the expression for $\delta^4 y_i$ as given by Eq. (5.11).
3. Obtain relations 5.22f and g for differential operators D^3 and D^4 in terms of backward difference operators.
4. Obtain relations 5.27d and e for differential operators D^3 and D^4 in terms of forward difference operators.
5. Obtain relations 5.40b and c for differential operators D^3 and D^4 in terms of central difference operators δ and μ.
6. Suggest a method for solving the differential equation

$$\frac{d^2 y}{dt^2} + y^2 = 0$$

with $dy/dt = 1$ and $y = 2$ at $t = 0$, using central differences of order of h^2.

7. Using the tabulated values of tan x for $x = 0.50$ to 0.60 radians in intervals of $\Delta x = 0.02$

x	tan x
0.50	0.54630
0.52	0.57256
0.54	0.59943
0.56	0.62695
0.58	0.65517
0.60	0.68414

find the values of $\tan x$ at $x = 0.55$ radian using the Gregory-Newton forward interpolation formula and Sterling's interpolation formula.

8. Obtain the first and second derivatives of $\tan x$ at $x = 0.54$ using the above table and relations given in Table 5-4, for errors of the order h^4. Compare these values to values obtained by differentiation of $\tan x$.

9. Using the tabulated values of $\tan x$ above, obtain the area under the tangent curve between $x = 0.5$ and $x = 0.58$ using trapezoidal and Simpson's one-third (central difference formulation) rules.

10. Take the equation

$$y = Ax^2 + Bx + C$$

and assume that it passes through the points $(0, y_i)$, $(\Delta x, y_{i+1})$ and $(2\Delta x, y_{i+2})$. Evaluate the coefficients A, B, and C and obtain

$$\text{Area} = \int_0^{2\Delta x} y \, dx$$

11. Same as the above problem, but this time find

$$\text{Area} = \int_{-\Delta x}^{3\Delta x} y \, dx$$

12. Assuming the polynomial

$$y = Ax^3 + Bx^2 + Cx + D$$

passing through $(0, y_i)$, $(\Delta x, y_{i+1})$, $(2\Delta x, y_{i+2})$, and $(3\Delta x, y_{i+3})$, find the values of A, B, C, and D. Show that the value of A is proportional to the third derivative of y with respect to x, $d^3 y/dx^3$. Compare the value of A with the forward difference Table 5-3.

REFERENCES

1. Boyce, William E., and Richard C. DiPrima: "Elementary Differential Equations," John Wiley & Sons, Inc., New York, 1969.

2. Sokolnikoff, I. S.: "Advanced Calculus," McGraw-Hill Book Com-
    pany, New York, 1939.
3. Pipes, Louis A.: "Applied Mathematics for Engineers and Physicists,"
    Chap. 10, McGraw-Hill Book Company, New York, 1958.
4. Kunz, K. S.: "Numerical Analysis," Chap. 4, McGraw-Hill Book
    Company, New York, 1957.
5. Hamming, Richard W.: "Numerical Methods for Scientists and
    Engineers," McGraw-Hill Book Company, New York, 1962.

6 NUMERICAL SOLUTION OF ORDINARY DIFFERENTIAL EQUATIONS

The numerical solution of differential equations is accomplished by writing the differential equation in finite-difference form in terms of discrete time steps (Δt) and solving the resulting difference equation. This solution of the finite-difference equation will depend on the chosen value of time step Δt. In the limit when the value of Δt approaches zero, the solution of the finite-difference equation will approach the solution of the differential equation. It should also be noted that the analytic solution of a differential equation gives the value of the dependent variable x as a function of the independent variable t. This function can be evaluated for any value of the independent variable. On the other hand, the finite-difference solution of a particular differential equation will represent a set of values of the dependent variable x as a function of the independent variable t. The x's will be obtained for values of independent variable $0, \Delta t, 2\Delta t, 3\Delta t, \ldots$, and so on. It will be seen that the solution x_n at $n \Delta t$, where n is an integer, will depend on the values of x for $(n - 1)\Delta t, (n - 2)\Delta t, \ldots$, and so on. This being the case, the starting values $x_0, x_1, x_2, \ldots$, should be calculated in order to be used in the general expression of the finite-difference representation.

METHODS OF STARTING THE SOLUTION

Taylor's Method

This method of starting the solution of a differential equation consists of representing the solution in the neighborhood of the initial values in a Taylor series expansion. This expansion will be an exact representation of the solution in the neighborhood of the initial conditions. Having this expansion, the values of the independent variable x can be evaluated for the few starting values of Δt.

Consider the first-order differential equation

$$x' = \frac{dx}{dt} = f(x, t) \tag{6.1}$$

with the initial conditions $x = x_0$ at $t = t_0$. Assuming the solution to be of the form

$$x = g(t) \tag{6.2}$$

we can write Taylor series expansion of x about $t = t_0$ as follows:

$$x = x_0 + (t - t_0)x_0' + \frac{1}{2!}(t - t_0)^2 x_0'' + \frac{1}{3!}(t - t_0)^3 x_0''' + \cdots \tag{6.3}$$

where primes denote derivatives with respect to t. These derivatives are evaluated at t_0. The first derivative is given by x' of Eq. (6.1), and the second derivative can be obtained by differentiating this with respect to t. Higher derivatives can be obtained in a similar fashion. That is,

$$x' = f(x, t) = g_1(x, t)$$

$$x'' = \frac{dx'}{dt} = \frac{\partial g_1}{\partial t}\frac{dt}{dt} + \frac{\partial g_1}{\partial x}\frac{dx}{dt} = \frac{\partial g_1}{\partial t} + \frac{\partial g_1}{\partial x}x' = g_2(t, x, x')$$

$$x''' = \frac{\partial g_2}{\partial x}x' + \frac{\partial g_2}{\partial t} + \frac{\partial g_2}{\partial x'}x'' = g_3(t, x, x', x'') \tag{6.4}$$

$$\vdots$$

$$x^s = g_s(t, x, x', x'', \ldots, x^{s-1})$$

where g's are functions of indicated variables. Starting with the first of Eq. (6.4), the values of x', x'', x''', ... are expressed as functions of preceding available values. For example, x'' is a function of t, x, and x', all of which are available from previous calculations. Note that in Eq (6.3) the values of derivatives are evaluated at the initial point x_0 and t around which the Taylor series expansion is written. Having the values given by Eq. (6.4), the series of Eq. (6.3) which represents the solution of Eq. (6.1) in the neighborhood of the initial condition can be written

Example. It is desired to evaluate the value of the function x expressed by the differential equation

$$\frac{dx}{dt} = t^3 + x \tag{6.5}$$

for $t = 1.1, 1.2, 1.3, 1.4,$ and 1.5. The initial conditions are $t = 1, x = 1$

Taylor series expansion corresponding to Eq. (6.3) can be written

$$x = 1 + (t - 1)x_1' + \frac{1}{2!}(t - 1)^2 x_1'' + \frac{1}{3!}(t - 1)^3 x_1''' + \cdots \tag{6.6}$$

where the values of the derivatives x', x'', ... are evaluated at $t_0 = 1$. The set of equations corresponding to Eq. (6.4) can be written

$$x_1' = (t^3 + x)_{(t=1, x=1)} = 2$$

$$x_1'' = (3t^2 + x')_{1,1} = 5$$

$$x_1''' = (6t + x'')_{1,1} = 11$$

$$x_1^{iv} = (6 + x''')_{1,1} = 17 \tag{6.7}$$

$$x^v = (x^{iv})_{1,1} = 17$$

$$x^{vi} = (x^v)_{1,1} = 17$$

Thus the Taylor series expansion corresponding to Eq. (6.6) becomes

$$x = 1 + (t - 1)(2) + \frac{1}{2}(t - 1)^2(5) + \frac{1}{6}(t - 1)^3(11) + \frac{1}{24}(t - 1)^4(17)$$

$$+ \frac{1}{120}(t - 1)^5(17) + \frac{1}{720}(t - 1)^6(17) + \cdots \tag{6.8}$$

The values of x can be evaluated using the terms given in Eq. (6.8) for the required values of t. These values are listed in Table 6-1.

The solution of the differential Eq. (6.5) can be obtained by the methods described in the Appendix. This solution is given:

$$x = Ae^t - t^3 - 3t^2 - 6t - 6 \tag{6.9}$$

where A is the constant to be evaluated from the initial condition $t = 1$, $x = 1$. This results in $A = 17/e$, or

$$x = 17e^{t-1} - t^3 - 3t^2 - 6t - 6 \tag{6.10}$$

The corresponding values of x for $t = 1.1, 1.2, \ldots$ are given in Table 6-1. Note that only the values of x obtained for $t = 1.5$ differ in the fifth decimal place.

Table 6-1. Required Values of x for the Example Problem of Eq. (6.5)

t	x	
	From Taylor series Eq. (6.8)	From solution Eq. (6.10)
1.0 (initial value)	1.00000	1.00000
1.1	1.22691	1.22691
1.2	1.51585	1.51585
1.3	1.88060	1.88060
1.4	2.33701	2.33701
1.5	2.90323	2.90326

Taylor Method for Higher-order and Sets of First-order Differential Equations

The above method can be directly applied to higher-order differential equations. Consider the nth-order differential equation

$$x^n = \frac{d^n x}{dt^n} = f(t, x, x', x'', \ldots, x^{n-1})$$

$$(6.11)$$

with the given initial conditions $x_0, x_0', x_0'', \ldots, x_0^{n-1}$ at t_0. The Taylor series for the variable x of this equation around t_0 can be written

$$x = x_0 + (t - t_0)x_0' + \frac{(t - t_0)}{2!}x_0'' + \frac{t - t_0}{3!}x_0''' + \cdots$$

The values of the derivatives to x_0^{n-1} are already available from the initial conditions. Additional derivatives as required in the Taylor series expansion can be obtained by using the initial values and Eq. (6.11).

In addition to this method, the nth-order differential equation (6.11) can be reduced to n first-order differential equations and Taylor's method of starting the solution applied to the set of these first-order equations. The reduction to n first-order equations is accomplished by defining new variables

$$x' = z_1(t, x)$$

$$z_1' = x'' = z_2(t, x, x') = z_2(t, x, z_1)$$

$$z_2' = x''' = z_3(t, x, z_1, z_2)$$

$$(6.12)$$

$$\vdots$$

$$z_{n-1}' = x^n = f(t, x, x', x'', \ldots, x^{n-1})$$

$$= f(t, x, z_1, z_2, \ldots, z_{n-1})$$

where $z_1, z_2, \ldots, z_{n-1}$ are newly defined variables. Thus the nth-order differential equation (6.11) is transformed to n first-order differential equations (6.12).

Taylor's method of starting the solution of a set of simultaneous differential equations is similar to the previously given procedure. To illustrate the application of Taylor's method to higher order and sets of simultaneous equations, consider the following examples.

Example. Consider the following second-order differential equation

$$x'' + x = 0, \quad x'' = \frac{d^2 x}{dt^2} \tag{6.13}$$

with the initial conditions $t = 0$, $x = 0$, $x' = 1$. The Taylor series representation of the solution is written

$$x = x_0 + t x_0' + \frac{t^2}{2!} x_0'' + \frac{t^3}{3!} x_0''' + \cdots \tag{6.14}$$

The values of derivatives are obtained from the initial conditions and Eq. (6.13) as follows.

$$
\begin{aligned}
x_0 &= 0 \\
x_0' &= 1 \\
x'' &= -x & x_0'' &= 0 \\
x''' &= -x' & x_0''' &= -1 \\
x^{\text{iv}} &= -x'' & x_0^{\text{iv}} &= 0 \\
x^{\text{v}} &= -x''' & x_0^{\text{v}} &= 1 \\
\vdots & & \vdots
\end{aligned} \tag{6.15}
$$

Thus the series of Eq. (6.14) becomes

$$x = t - \frac{t^3}{3!} + \frac{t^5}{5!} - \frac{t^7}{7!} + \cdots \tag{6.16}$$

This solution can also be obtained by writing Eq. (6.13) as two first-order differential equations

$$
\begin{aligned}
x' &= z \\
z' &= -x
\end{aligned} \tag{6.17}
$$

and writing Taylor series expansion for x and z. This solution, in addition to the values of the function x, will give its derivative z. Of course, once the function in the form of series (6.16) has been obtained, the derivative can be obtained merely by differentiation of Eq. (6.16). The series of Eq. (6.16) can be recognized as the series for $\sin t$. Thus, the values of can be obtained for $t = 0.05, 0.10, \ldots$ using the first three terms of the series as given in Table 6-2.

Differential equation (6.13) is recognized as the second-order differential equation representing the harmonic motion. The solution of this equation from elementary differential equation or the Appendix is

$$x = A \sin t + B \cos t \qquad (6.18$$

Using the initial condition $t = 0$, $x = 0$, $x' = 1$, this solution becomes

$$x = \sin t \qquad (6.19$$

which is the solution represented by the series of Eq. (6.17). The values of x evaluated using Eq. (6.19) are also given in Table 6-2.

Example. Consider Taylor series expansion of the solution of the simultaneous system of differential equations

$$\frac{dx}{dt} = x^2 + t$$

$$\frac{d^2 y}{dt^2} = x + y - t \qquad (6.20$$

Table 6-2. Required Values of x for the Example Problem of Eq. (6.13)

t	x	
	Three terms of series Eq. (6.16)	Solution Eq. (6.19)
0 (initial condition)	0.00000	0.00000
0.05	0.04998	0.04998
0.10	0.09983	0.09983
0.15	0.14944	0.14944
0.20	0.19867	0.19867
0.25	0.24740	0.24740

with the initial conditions $t = 0$, $x = 1$, $y = 0$, $y' = 1$. Note that the first equation of the set is nonlinear as x appears to the second power. Equations (6.20) can be written as a set of three first-order differential equations by introducing z as the additional variable. This results:

$$x' = x^2 + t$$
$$y' = z \qquad\qquad (6.21)$$
$$z' = x + y - t$$

with the initial conditions $t = 0$, $x = 1$, $y = 0$, $z = 1$. The solutions expressed in Taylor series expansion about $t = 0$ can be written

$$x = x_0 + tx_0' + \frac{t^2}{2!} x_0'' + \frac{t^3}{3!} x_0''' + \cdots$$

$$y = y_0 + ty_0' + \frac{t^2}{2!} y_0'' + \frac{t^3}{3!} y_0''' + \cdots \qquad (6.22)$$

$$z = z_0 + tz_0' + \frac{t^2}{2!} z_0'' + \frac{t^3}{3!} z_0''' + \cdots$$

The values of the derivatives at $t = 0$ can be obtained by starting with Eqs. (6.21) as follows:

$$\begin{cases} x_0' = (x^2 + t)_{t=0} = 1 \\ y_0' = (z)_{t=0} = 1 \\ z_0' = (x + y - t)_{t=0} = 1 \end{cases}$$

$$\begin{cases} x_0'' = (2xx' + 1)_{t=0} = 3 \\ y_0'' = (z')_{t=0} = 1 \\ z_0'' = (x' + y' - 1)_{t=0} = 1 \end{cases}$$

(continued)

$$\begin{cases} x_0''' = [2(x')^2 + 2xx'']_{t=0} = 8 \\ y_0''' = (z'')_{t=0} = 1 \\ z_0''' = (x'' + y'')_{t=0} = 4 \end{cases}$$

(6.23

$$\begin{cases} x_0^{iv} = (4x'x'' + 2x'x'' + 2xx''')_{t=0} = 34 \\ y_0^{iv} = (z''')_{t=0} = 4 \\ z_0^{iv} = (x''' + y''')_{t=0} = 9 \end{cases}$$

Using the above values, the series of Eq. (6.22) becomes

$$x = 1 + t + \frac{3}{2}t^2 + \frac{4}{3}t^3 + \frac{17}{12}t^4 + \cdots$$

$$y = t + \frac{t^2}{2} + \frac{t^3}{6} + \frac{t^4}{6} + \cdots$$

(6.24

$$z = 1 + t + \frac{t^2}{2} + \frac{2}{3}t^3 + \frac{3}{8}t^4 + \cdots$$

The initial values of x, y, and z can be evaluated from the above series.

Euler Method of Starting the Solution

The Euler method of starting the solution of a differential equation in effect uses the definition of the derivative as the ratio of two discrete intervals and proceeds to calculate values in the neighborhood of the initial starting point. This method, in many applications, is also used as a method for continuing the solutions. Consider the differential equation

$$\frac{dy}{dx} = f(x, y)$$

(6.25

with the initial conditions $x = x_0$ and $y = y_0$. This equation can be written using the definition of derivative

$$\frac{dy}{dx} = \lim_{\Delta x \to 0} \frac{\Delta y}{\Delta x}$$

(6.26

For small Δx's

$$\Delta y \simeq \frac{dy}{dx} \Delta x \tag{6.27}$$

$$\simeq f(x, y) \Delta x$$

Thus, assuming a Δx and using the function $f(x, y)$ from Eq. (6.25), the value of Δy can be calculated. The value of y for the next increment in x will be $y + \Delta y$. Although this method of computation is used for starting solutions, better results are obtained by using Taylor's method. Note that the increment Δx in Euler's method can be made small to improve the accuracy of the solution.

Example. Consider the differential equation

$$\frac{dy}{dx} = \frac{x^2 - y}{1 + y} \tag{6.28}$$

with the initial conditions $x = 0$, $y = 1$. It is desired to find the values of y for $x = 0.1$ and 0.2 by Euler's method. Let us take a Δx increment of 0.02 and write Eq. (6.28) in notation (6.27)

$$\Delta y = \left(\frac{x^2 - y}{1 + y}\right)_{\substack{x = 0 \\ y = 1}} (0.02) \tag{6.29}$$

$$= -0.01$$

Thus the value of y for $x = 0.02$ will be $y = 1 - 0.01 = 0.99$. We continue the solution by writing Eq. (6.29) for $x = 0.02$ and $y = 0.99$ as

$$\Delta y = \left(\frac{x^2 - y}{1 + y}\right)_{\substack{x = 0.02 \\ y = 0.99}} (0.02)$$

$$= -\left(\frac{0.9896}{1.99}\right)(0.02) \tag{6.30}$$

$$= -0.0099457$$

and the value of y becomes

$$y = 0.99 - 0.0099457 = 0.980054$$

Continuing the above operation for $x = 0.04, 0.06, \ldots$ we obtain values of y given in Table 6-3. From this table the values of y for $x = 0.1$ and 0.2 can be obtained as required in the problem. Note that by assuming $\Delta x = 0.1$, Eq. (6.29) can be solved for Δy and, subsequently, y. This results in

$$\Delta y = \left(\frac{x^2 - y}{1 + y}\right)_{\substack{x = 0 \\ y = 1}} (0.1) = -0.05 \tag{6.31}$$

and

$$y = 1.0 - 0.05 = 0.95 \tag{6.32}$$

This value differs from the more accurate value of y for $x = 0.1$ as given in Table 6-3. Continuing the operation with these values and the equation

$$\Delta y = \left(\frac{x^2 - y}{1 + y}\right)_{\substack{x = 0.1 \\ y = 0.95}} (0.1) \tag{6.33}$$

Table 6-3.   Values of y as a Function of x (Eq. (6.28))

x	y
0.00	1.000000
0.02	0.990000
0.04	0.980054
0.06	0.970171
0.08	0.960359
0.10	0.950627
0.12	0.940982
0.14	0.931435
0.16	0.921993
0.18	0.912665
0.20	0.903461

the value of y at $x = 0.2$ becomes

$$y = 0.95 + \Delta y = 0.901795$$

which again differs from the corresponding value given in Table 6-3.

Note that if the differential equation contains fractions of the variables as in Eq. (6.28), the application of Euber's method for starting the solution is simpler than Taylor's expansion. The reason for this is that Taylor's expansion requires first- and higher-order derivatives which are tedious to obtain in case of fractions.

METHODS OF CONTINUING SOLUTION OF DIFFERENTIAL EQUATIONS

Having obtained several starting values of the dependent variable, we can use the general finite difference expressions for continuing the solution. The values of time or space increments, Δt and Δx, can be changed during the methods of continuing the solution. For example, if during the initial transients, a higher solution accuracy is required, a smaller increment can be used and the increment increased during the latter portions of the solution. In this chapter, we will discuss the methods of continuing the solution for initial value problems while in the next chapter the method of solutions of boundary value problems will be discussed.

The *initial value* problems are the ones in which the values of the dependent variable and its derivatives are specified at an initial point of the independent variable (these problems are also referred to as *one-point boundary value* problems). The initial point refers to the start of the problem. For example, in the differential equation

$$\frac{d^2 x}{dt^2} + x = 0 \tag{6.34}$$

the initial conditions can be expressed as the values x and dx/dt at some value of $t = t_0$. The solution of the differential equation will then apply for values of $t > t_0$. Since Eq. (6.34) is a second-order differential equation, two conditions will be required to specify the constants of the solution. In the above case, the values of x and its derivative constituted these conditions. This is to be compared to boundary value problems where the conditions at various values of the independent variable are given. In the case of Eq. (6.34) if the values of x at t_0 and x at $t_1 > t_0$ were specified, again the constants of the solution of the differential equation can be evaluated. Specifying these values converts the problem to a boundary value problem (also referred to as *two-point boundary value* problems).

Most of the problems in electrical engineering are initial value problems as the independent variable of the differential equation is time. In these problems, conditions at time equal to zero are given and the solution is required for $t > 0$. However, in mechanical engineering, boundary value problems are encountered in elasticity and heat transfer. Examples of these will be bending problems in strength of material where the conditions at the end points are specified.

In this chapter we will discuss the numerical solution of initial value problems, and in the next chapter we will extend these techniques to boundary value problems.

Milne's Method of Solution of Initial Value Problems

Consider the differential equation

$$y' = \frac{dy}{dx} = f(x, y) \tag{6.35}$$

and assume that the first four points y_0, y_1, y_2, y_3 corresponding to x_0, x_1, x_2, x_3 are calculated by the previously given methods of starting the solution. Denoting the interval between x's as $\Delta x = x_1 - x_0 = x_2 - x_1 = x_3 - x_2 = \cdots$ Eq. (5.71) of Chapter 5 can be written

$$\int_{-2\Delta x}^{2\Delta x} y' \, dx = \frac{4\Delta x}{3} (2f_{i+1} - f_i + 2f_{i-1}) \tag{6.36}$$

Note that in the above expression $y' = f(x, y)$ is substituted for y of Eq. (5.71). The left side of Eq. (6.36) can be written

$$\int_{-2\Delta x}^{2\Delta x} \frac{dy}{dx} \, dx = \int_{y_{i-2}}^{y_{i+2}} dy = y_{i+2} - y_{i-2} \tag{6.37}$$

where i is used as an index. Using Eq. (6.37) in Eq. (6.36), we get

$$y_{i+2} - y_{i-2} = \frac{4\Delta x}{3}(2f_{i+1} - f_i + 2f_{i-1}) \tag{6.38}$$

It is more desirable to write the expression for y_{i+1} in terms of previous y's. This can be obtained by merely setting the index $i = i - 1$ in the above expression:

$$y_{i+1} = y_{i-3} + \frac{4\Delta x}{3}(2f_i - f_{i-1} + 2f_{i-2}) \tag{6.39}$$

Having the values of x_0 through x_3 and the corresponding values of y_0 through y_3, the values of f_0 through f_3 can be calculated by substitution in $f(x, y)$ of Eq. (6.35). Then by setting $i = 3, 4, 5, 6, \ldots$ in Eq. (6.39), the values of $y_4, y_5, \ldots$ for $x_4 = x_3 + \Delta x$, $x_5 = x_4 + \Delta x$, $\ldots$ can be calculated. Equation (6.39) is referred to as Milne's *predictor* formula.

In some problems it may be desirable to check or correct the solution by another integration formula. This is accomplished as follows. From Eq. (6.39) we obtain the value of y_{i+1} corresponding to x_{i+1}. Setting these values in Eq. (6.35) will give f_{i+1}. Equation (5.70) of Chapter 5 can be written in the form

$$\int_{-\Delta x}^{\Delta x} y' \, dx = \frac{\Delta x}{3}(y'_{i+1} + 4y'_i + y'_{i-1})$$

where y is replaced by y'. Replacing y' on the right of the above equation by f, we get

$$y_{i+1} = y_{i-1} + \frac{\Delta x}{3}(f_{i-1} + 4f_i + f_{i+1}) \tag{6.40}$$

Equation (6.40) is termed the *corrector* equation. The value of y_{i+1} obtained from Eq. (6.40) is substituted in Eq. (6.35) and a new f_{i+1} is obtained. This value is then substituted in Eq. (6.40) and a second iteration of y_{i+1} is calculated. If this y_{i+1} and the previous y_{i+1} agree to some specified accuracy, then y_{i+2} is calculated from Eq. (6.39); if not, the iteration given by Eq. (6.40) is repeated. However, in practice, a single application of corrector Eq. (6.40) is sufficient. It should be pointed out that in some applications the *predictor-corrector* method described above gives unstable solutions (see [2]). In addition, as seen from Eq. (6.39), Milne's method requires four initial values computed by the previously given methods for starting the solution.

Example. Consider the previously given differential equation (6.5)

$$\frac{dx}{dt} = t^3 + x = f(x, t) \tag{6.41}$$

with the initial condition $t = 1, x = 1$. Assuming that the first four values of x as a function t are computed (Taylor series solution of Table 6-1), the value of $f(x, t)$ in Eq. (6.41) can be obtained as follows:

$$t_0 = 1.0 \qquad x_0 = 1.00000 \qquad f(x_0, t_0) = f_0 = 2.00000$$

$$t_1 = 1.1 \qquad x_1 = 1.22691 \qquad\qquad\qquad f_1 = 2.55791$$

$$t_2 = 1.2 \qquad x_2 = 1.51585 \qquad\qquad\qquad f_2 = 3.24385$$

$$t_3 = 1.3 \qquad x_3 = 1.88060 \qquad\qquad\qquad f_3 = 4.07760$$

Having these values, we can apply the predictor Eq. (6.39) with $i = 3$ and calculate x_4 (same as y_4 of Eq. (6.39))

$$x_4 = x_0 + \frac{4\Delta t}{3}(2f_3 - f_2 + 2f_1) \tag{6.42}$$

Assuming that $\Delta t = 0.1$, this becomes

$$x_4 = 1.0 + \left(4\frac{0.1}{3}\right)(2 \times 4.07760 - 3.24385 + 2 \times 2.55791)$$

$$= 2.33696 \tag{6.43}$$

which is the predicted value of x for $t_4 = t_3 + \Delta t = 1.3 + 0.1 = 1.4$. Setting this value of x_4 in Eq. (6.41), we get

$$f_4(x_4, t_4) = (1.4)^3 + 2.33696 = 5.08096 \tag{6.44}$$

Using this value of f_4 in the corrector equation (6.40) with $i = 3$, we get

$$x_4 = x_2 + \frac{\Delta t}{3}(f_2 + 4f_3 + f_4) = 1.51585 + \frac{0.1}{3}(3.24385 + 4$$

$$\times 4.07760 + 5.08096) = 2.33702 \qquad (6.45)$$

which is the corrected value of x_4. Assuming that a single application of the corrector formula is sufficient (x_4 in Eq. (6.45) is close to x_4 in Eq. (6.43)), we calculate f_4 with the corrected value of x_4

$$f_4 = (1.4)^3 + 2.33702 = 5.08102 \qquad (6.46)$$

Now we return to the predictor equation (6.39) with $i = 4$ and calculate the value of x_5. From this point the process is repeated, resulting in the values given in Table 6-4. In this table time represents the values of t,

Table 6-4. Solution of the Example Problem Eq. (6.41)

TIME	PRED X (Eq. (6.43))	CORR X (Eq. (6.45))	FUNCTION (Eq. (6.46))	ANALT SOL (Eq. (6.10))
1.0	1	1	2	1
1.1	1.22691	1.22691	2.55791	1.22691
1.2	1.51585	1.51585	3.24385	1.51585
1.3	1.88060	1.88060	4.07760	1.88060
1.4	2.33696	2.33702	5.08102	2.33702
1.5	2.90320	2.90326	6.27826	2.90326
1.6	3.59994	3.60002	7.69602	3.60002
1.7	4.45071	4.45080	9.36380	4.45080
1.8	5.48210	5.48220	11.3142	5.48220
1.9	6.72415	6.72426	13.5833	6.72425
2.0	8.21068	8.21080	16.2108	8.21079
2.1	9.97970	9.97983	19.2408	9.97982
2.2	12.0739	12.0740	22.7220	12.0740
2.3	14.5409	14.5411	26.7081	14.5410
2.4	17.4342	17.4344	31.2584	17.4344
2.5	20.8135	20.8137	36.4387	20.8137
2.6	24.7454	24.7456	42.3216	24.7456
2.7	29.3039	29.3041	48.9871	29.3041
2.8	34.5718	34.5720	56.5240	34.5720
2.9	40.6409	40.6412	65.0302	40.6412
3.0	47.6137	47.6140	74.6140	47.6140
3.1	55.6036	55.6039	85.3949	55.6039

PRED X refers to values obtained from Eq. (6.42), CORR X refers to values obtained from Eq. (6.45), and FUNCTION refers to values obtained from Eq. (6.46). The analytic solution which is the value of the solution given by Eq. (6.10) is listed in the last column of Table 6-4. As seen from the table the agreement between the analytical and finite difference solutions is excellent.

Digital Computer Program for the Above Example

The following digital computer program has been written in the BASIC computer language for a General Electric time-sharing computer. Although most of the statements of the program are self-explanatory, for further detail about this computer language the reader is referred to [3].

```
4     DIM T(100),X(100),Z(100),F(100)
5     DEF FNF(T)=17*EXP(T-1)-T↑3-3*(T↑2)-6*T-6
10    PRINT "TIME,""PRED X," "CORR X," "FUNCTION," "ANALT SOL"
11    PRINT
20    LET T(0)=1.
30    LET T(1)=1.1
40    LET T(2)=1.2
50    LET T(3)=1.3
60    LET X(0)=1.
70    LET X(1)=1.22691
80    LET X(2)=1.51585
90    LET X(3)=1.88060
95    FOR I=0 TO 3
100   LET F(I)=(T(I))↑3+X(I)
105   LET Z(I)=X(I)
106   PRINT T(I),X(I),Z(I),F(I),FNF(T(I))
110   NEXT I
112   FOR I=3 TO 20
115   LET T(I+1)=T(I)+.1
120   LET X(I+1)=X(I-3)+((4*.1)/3)*(2*F(I)-F(I-1)+2*F(I-2))
130   LET F(I+1)=(T(I+1))↑3+X(I+1)
140   LET Z(I+1)=Z(I-1)+(0.1/3)*(F(I-1)+4*F(I)+F(I+1))
145   LET F(I+1)=(T(I+1))↑3+Z(I+1)
150   PRINT T(I+1),X(I+1),Z(I+1),F(I+1),FNF(T(I+1))
155   LET X(I+1)=Z(I+1)
160   NEXT I
170   END
```

The program starts with the dimension statement 4, reserving computer memory locations for time T predicted x, X, corrected x, Z, and the value of the function $f(x, t), F$.   Statement 5 defines the function FNF (T), which represents the analytic solution of the problem given by Eq. (6.10). Statements 20 through 90 enter the four initial conditions. Statements 95 through 110 calculate the values of the function for these initial values, and in addition set the corrected values of x and the predicted values of x for the four initial points equal, i.e., statement 105. Statements 112 through 160 calculate the answers. Statement 120 is the predictor equation (6.39) and statement 140 is the corrector equation (6.40).   Note that after values are printed out by statement 150 the values of corrected x's are substituted for predicted values—statement 155.

The program computes values for 20 time steps (statement 112), as given in Table 6-4.

Stability of Solutions

In numerical computations, problems are formulated in terms of finding the values of dependent variable y_i, in terms of previously computed values of this variable, its derivatives, the independent variable, and the increment h or Δt.   Denoting the available values of variables by x_1 through x_n, the problem can be written in equation form

$$y_i = f_i(x_1, x_2, \ldots, x_n, h) \tag{6.47}$$

Since the previously computed values of x's contain errors that may have been caused by, for example, truncation of significant figures, or the use of a particular difference scheme, the error will be reflected in succeeding values of y's approximately according to the relation

$$\delta y_i = \sum_{j=1}^n \frac{\partial f_i}{\partial x_j} \delta x_j \tag{6.48}$$

where δy_i and δx_j are errors in the computed values of y_i and x_j. If the coefficients of δx_j terms, $\partial f_i/\partial x_j$, are reasonably small, then δx_j errors will not degrade the solutions obtained for y_i. In this case the solutions are termed *stable*. Note that these coefficients are also a function of the increment h. On the other hand, if the coefficients $\delta f_i/\delta x_j$ are large, then seemingly small errors in δx_j values are magnified by these coefficients and repeated iterations will increase the error in y_i values, rendering the solution meaningless. Under these conditions the solutions are termed unstable. Note that Eq. (6.47) represents the particular finite-difference formulation of a problem and that the terms of Eq. (6.48) are a function of this representation. For this reason in a number of cases the actual solution of the problem may be stable but its finite-difference representation may result in an unstable solution. Analytical methods of determining the stability of a given finite difference representation will be discussed in the next chapter.

Since the digital computers presently in operation carry a great number of significant figures, the computational errors and thereby the stability problems resulting from computations are usually small. In many engineering problems the inherent stability of finite-difference formulations is discovered when the problems are run on the computers and the solutions show behavior incompatible with physical and engineering reasoning. This method of analysis, of course, has no theoretical justification, but it reveals possible stability problems with the least amount of effort; theoretical analysis of stability criteria is extremely time consuming. Furthermore, in engineering applications the solutions are desired primarily during the initial transients, and most solutions that become unstable as the values of independent variable increases may be stable in the region of interest.

To illustrate the above discussion we will analyze the errors in the finite difference formulation of the following example problem.

Example. Consider the solution of the previously given example of Eq. (6.41). It is seen from Table 6-4 that the evaluated solution gives perfectly acceptable answers for the range of values of $1 \leq T \leq 3.1$. Table 6-4a gives the values evaluated to $T = 10.0$ under the identical conditions of time step and initial values. From this table it is noted that as T assumes larger values (although the solution remains acceptable), the error in the solution increases. The error is defined as the difference between the value given by the corrector formula and the analytical values obtained.

From Table 6-4a it is seen that the percentage error (e.g., at $T = 10.0$ the percent error is $(1.75/136386) \times 100 = 1.3 \times 10^{-3}$) remains very small. As discussed above, the error is partially caused by truncation of significant figures in the four initial values given following Eq. (6.41).

As mentioned before, the stability of solution of predictor corrector methods has been studied extensively, as given in [2, 4]. In addition, [4] suggests a corrector formula developed by R. W. Hamming to replace

Table 6-4a. Error in the Computed Values Using Predictor-Corrector Method

Time T	Value of solution		Absolute value of error*
	Corrector formula	Analytic Solution	
1.0	1.0	1.0	0
2.0	8.21084	8.21079	4.46×10^{-5}
3.0	47.6143	47.6140	3.03×10^{-4}
4.0	199.455	199.454	1.32×10^{-3}
5.0	692.173	692.168	4.94×10^{-3}
6.0	2157.04	2157.02	1.71×10^{-2}
7.0	6320.34	6320.29	5.68×10^{-2}
8.0	17884.9	17884.8	1.82×10^{-1}
9.0	49644.8	49644.3	5.69×10^{-1}
10.0	136388.0	136386.0	1.75

*The values of errors were computed in the computer, which carries more significant figures than shown in the printout of the answers.

Eq. (6.40). This formula in the previously given notation is as follows

$$y_{i+1} = \frac{1}{8}\left[9y_i - y_{i-2} + 3\Delta x(f_{i+1} + 2f_i - f_{i-1})\right] \tag{6.49}$$

In [4] the errors developed using this equation and Eq. (6.40) are analyzed in the case of a particular example. It is shown that for this example the error in the solution using Eq. (6.49) as a corrector is less than using Eq. (6.40), which is Milne's corrector equation.

Adams' Method

Adams' method of solving differential equations starts with the expression for backward difference of a function in terms of differentials of the function, as given by Eq. (5.22a) of Chapter 5

$$\nabla y_{i+1} = \left(1 - \frac{hD}{2} + \frac{h^2 D^2}{6} - \frac{h^3 D^3}{24} + \cdots\right) h Dy_i \tag{6.50}$$

where h is the increment of the independent variable. Substituting the values of differential operators D in terms of backward difference operators (Eqs. (5.22)), we get

$$y_{i+1} = y_i + h\left(1 + \frac{\nabla}{2} + \frac{5}{12}\nabla^2 + \frac{3}{8}\nabla^3 + \cdots\right)Dy_i \qquad (6.51)$$

In obtaining the above expression the relation

$$\nabla y_{i+1} = y_{i+1} - y_i$$

is used. Note that in a given differential equation

$$\frac{dy}{dx} = f(x, y) \qquad (6.52)$$

the value of $Dy_i = dy_i/dx = f_i$. Using this, Eq. (6.51) becomes

$$y_{i+1} = y_i + h\left(1 + \frac{\nabla}{2} + \frac{5}{12}\nabla^2 + \frac{3}{8}\nabla^3 + \cdots\right)f_i \qquad (6.53)$$

This relation is Adams' recursion equation for the solution of differential equations. Note that the values of y are based on the backward differences of f_i.

In using the above relations let us assume that four initial values of y for x_0 through x_4 and the corresponding values of f_i are calculated. That is, we have

x_0	y_0	f_0			
x_1	y_1	f_1	∇f_1		
x_2	y_2	f_2	∇f_2	$\nabla^2 f_2$	
x_3	y_3	f_3	∇f_3	$\nabla^2 f_3$	$\nabla^3 f_3$

where the differences (∇'s) of f_i are obtained from the given values. Having these values, y_4 can be computed by setting $i = 3$ in Eq. (6.53)

$$y_4 = y_3 + h\left(f_3 + \frac{\nabla f_3}{2} + \frac{5}{12}\nabla^2 f_3 + \frac{3}{8}\nabla^3 f_3\right) \qquad (6.54)$$

After obtaining y_4 another row in the above table can be computed and iteration can proceed to y_5, y_6, and so on. Adams' method also requires several starting values to continue the computations.

Fox-Euler Method

This method can be considered an extension of Euler's method of starting the solution of differential equations as described previously and depicted in Eq. (6.27). This equation can also be obtained from the Taylor series expansion

$$y_{i+1} = y_i + hy_i' + \frac{h^2}{2!} y_i'' + \frac{h^3}{3!} y_i''' + \cdots \tag{6.55}$$

by neglecting terms involving second-order and higher derivatives. This results in

$$y_{i+1} - y_i = hy' = h\frac{dy}{dx} \tag{6.56}$$

which is identical to Eq. (6.27) with $\Delta x = h$. Suppose, in expression (6.55), instead of neglecting the second-order derivative term we substitute this term with the value of y_i'' in terms of forward differences of y', that is,

$$y_i'' = \frac{y_{i+1}' - y_i'}{h} \tag{6.57}$$

This will result:

$$y_{i+1} = y_i + hy_i' + h\frac{y_{i+1}' - y_i'}{2} \tag{6.58}$$

Or, finally, we get

$$y_{i+1} = y_i + \frac{h}{2}(y_{i+1}' + y_i') \tag{6.59}$$

which is the Fox-Euler integration formula. Note that the second term on the right of the above equation is the average value of the derivative between y_{i+1} and y_i multiplied by h. Equation (6.59) can be used without the application of Taylor series for obtaining the starting values.

Example. Consider the solution of the differential equation

$$\frac{dy}{dx} = x^2 + y = y' \tag{6.60}$$

with the initial conditions $x = 0$, $y = 1$. The analytical solution of this equation using the initial conditions is

$$y = 3e^x - x^2 - 2x - 2 \tag{6.61}$$

Using Eq. (6.60) in Eq. (6.59) results in

$$y_{i+1} = y_i + \frac{h}{2}\left(x_{i+1}^2 + y_{i+1} + x_i^2 + y_i\right) \tag{6.62}$$

Solving this equation for y_{i+1}, we get

$$y_{i+1} = \frac{1}{1-(h/2)}\left[y_i + \frac{h}{2}\left(x_{i+1}^2 + x_i^2 + y_i\right)\right] \tag{6.63}$$

By setting the values of $i = 0, 1, 2, 3, \ldots$ and assuming $y = 1$ for $x_0 = 0$, the above solution can be evaluated for increments $\Delta x = h$ of x. This solution is given in Table 6-5 for $h = 0.5$.

For comparison, the analytical solution which is the evaluation of Eq. (6.61) is also shown.

Using Eq. (6.56), which is Euler's method for starting the solution, the values of y can also be evaluated as a function of x. This will give

$$y_{i+1} = y_i + hy_i' \tag{6.64}$$

which for the case of the example becomes

$$y_{i+1} = y_i + h\left(x_i^2 + y_i\right)$$
$$= y_i(1 + h) + hx_i^2 \tag{6.65}$$

Table 6-5. Solution of Eq. (6.60)

x	y (Eq. (6.63))	y (Eq. (6.65))	y (Eq. (6.61))
)	1	1	1
05	1.05135	1.05	1.05131
1	1.10558	1.10263	1.10551
15	1.16311	1.15826	1.163
2	1.22436	1.21729	1.22421
25	1.28978	1.28016	1.28958
3	1.35983	1.34729	1.35958
35	1.43501	1.41916	1.4347
4	1.51585	1.49624	1.51547
45	1.60288	1.57905	1.60244
5	1.69668	1.66813	1.69616
55	1.79786	1.76404	1.79726
6	1.90704	1.86736	1.90636
65	2.0249	1.97873	2.02412
7	2.15214	2.09879	2.15126
75	2.28949	2.22823	2.2885
8	2.43774	2.36777	2.43662
85	2.59768	2.51816	2.59644
9	2.77019	2.68019	2.76881
95	2.95616	2.8547	2.95463
1.	3.15655	3.04256	3.15485

The solution resulting from this equation is also shown for comparison in Table 6-5. From this table it is noted that the values of y obtained from Eq. (6.63) more closely approximate the theoretical values of Eq. (6.61). The simpler equation (6.65) will give more accurate results if the time step is further decreased to, for example, 0.025.

Solution of Simultaneous First-order Differential Equations

The governing principles of the above solutions can equally well be applied to series of simultaneous first-order differential equations. As discussed previously and depicted by Eqs. (6.12), higher-order differential equations can be written as a set of first-order simultaneous differential equations. Thereby, the methods of solution of simultaneous first-order differential equations is useful in solving higher-order equations.

Consider the set of first-order differential equations

$$y' = f(x, y, z)$$
$$z' = \varphi(x, y, z)$$

$$(6.66)$$

where the derivatives are with respect to x and the initial conditions are $x = x_0$, $y = y_0$, and $z = z_0$. Application of Adams' method through Eq. (6.53) will result in

$$y_{i+1} = y_i + h\left(1 + \frac{\nabla}{2} + \frac{5}{12}\nabla^2 + \cdots\right)f_i$$

$$z_{i+1} = z_i + h\left(1 + \frac{\nabla}{2} + \frac{5}{12}\nabla^2 + \cdots\right)\varphi_i$$

$$(6.67)$$

where h is the increment in the independent variable x_0. After obtaining the first few values of y and z for initial x's by Taylor's method of starting the solution, Eqs. (6.67) can be applied much in the same way as for a single first-order differential equation.

Example. Consider the previously given example problem of Eq. (6.21):

$$x' = x^2 + t = f(x, t)$$
$$y' = z = g(z)$$
$$z' = x + y - t = \varphi(x, y, t)$$

$$(6.68)$$

where primes indicate derivatives with respect to t. Assume that the following initial values are given

$$t = t_0, x_0, y_0, z_0, f_0, g_0, \varphi_0$$

$$(6.69)$$

Neglecting terms involving ∇ and ∇^2 in Eqs. (6.67), these equations can be written for x, y, and z as follows:

$$x_{i+1} = x_i + hf_i$$
$$y_{i+1} = y_i + hg_i$$
$$z_{i+1} = z_i + h\varphi_i$$

$$(6.70)$$

Using conditions given in Eq. (6.69), the above equations can be solved for $i = 0$, resulting in x_1, y_1, and z_1. Having these values, f_1, g_1, and φ_1 can be calculated from Eq. (6.68). Then the application of Eq. (6.70) for $i = 1$ will result in the values of x_2, y_2, and z_2. The solution can then proceed as before.

Note that neglecting terms involving ∇ and ∇^2 in Eq. (6.67) reduced this formula to Euler's relation (6.27).

Milne's Method for Second-order Equations

In addition to the method of solution described above, in which a second-order differential equation can be reduced to two first-order differential equations and then solved, methods exist for the direct solution of second-order differential equations. The first method to be described is that of Milne. This method is similar to the previously described Milne's method for first-order differential equations.

Consider the second-order differential equation

$$y'' = f(x, y, y') \tag{6.71}$$

where primes indicate derivatives with respect to x, with the initial conditions $x = x_0$, $y = y_0$, and $y' = y_0'$. The values of y_1, y_1', y_2, y_2', y_3, y_3' corresponding to x_1, x_2, and x_3 can be obtained using Taylor series expansion described for starting the problem. Having these initial values, f_0, f_1, f_2, and f_3 can be computed from Eq. (6.71).

Using Eq. (5.71) of Chapter 5 for y'' instead of y, we get

$$\int_{-2\Delta x}^{2\Delta x} y'' \, dx = \frac{4\Delta x}{3} (2f_{i+1} - f_i + 2f_{i-1}) \tag{6.72}$$

where the function of Eq. (6.71) is used in the right of this equation. Since

$$\int_{-2\Delta x}^{2\Delta x} y'' \, dx = \int_{-2\Delta x}^{2\Delta x} \frac{dy'}{dx} \, dx = \int_{y_{i-2}'}^{y_{i+2}'} dy' = y_{i+2}' - y_{i-2}' \tag{6.73}$$

Eq. (6.72) can be written

$$y'_{i+2} = y'_{i-2} + \frac{4\Delta x}{3}(2f_{i+1} - f_i + 2f_{i-1}) \tag{6.74}$$

Setting the value of index $i = i - 1$, Eq. (6.74) can be written in a more convenient form as

$$y'_{i+1} = y'_{i-3} + \frac{4\Delta x}{3}(2f_i - f_{i-1} + 2f_{i-2}) \tag{6.75}$$

The above equation gives the value of y'_{i+1} in terms of four previous values of y' and f. Having obtained the value of y'_{i+1} from Eq. (6.75), Eq. (5.70) of Chapter 5, Simpson's one-third rule, gives

$$y_{i+1} = y_{i-1} + \frac{\Delta x}{3}(y'_{i+1} + 4y'_i + y'_{i-1}) \tag{6.76}$$

At this point we have all of the required values at $i + 1$ step, that is, $x_{i+1} = x_i + \Delta x$, y_{i+1} from Eq. (6.76) and y'_{i+1} from Eq. (6.75). Having these values, f_{i+1} can be computed from Eq. (6.71).

At this point the corrector formula

$$y'_{i+1} = y'_{i-1} + \frac{\Delta x}{3}(f_{i+1} + 4f_i + f_{i-1}) \tag{6.77}$$

can be applied to obtain a better approximation for y'_{i+1} than the previous value as obtained from Eq. (6.75). Note that the primary difference between values of y'_{i+1} obtained from Eqs. (6.75) and (6.77) lies in the fact that in Eq. (6.75) the value of y'_{i+1} is based on the values of f up to ith increment while in Eq. (6.77) the value of y'_{i+1} is based on the prior values of f as well as the current value, f_{i+1}. Therefore it is believed that Eq. (6.77) will give a better value for y'_{i+1}. Having this value of y'_{i+1}, Eq. (6.76) is reapplied and the value of y_{i+1} computed. The procedure outlined following this equation may be reapplied to compute another value of y'_{i+1} from Eq. (6.77), but in most cases a single application of the corrector formula (6.77) is sufficient. For the stability discussion of predictor-corrector methods the reader is referred to [4].

Noumerov's Method of Solving Second-order Differential Equations

When the first derivative term of a second-order differential equation is missing, Noumerov's method of solution can be applied. Consider the following differential equation with derivatives with respect to x

$$y'' + f(x)y = F(x) \tag{6.78}$$

with the y' term absent. Substitution of y'' in terms of its central difference representation as given in Eq. (5.40a) of Chapter 5 will result in

$$\left(\delta^2 - \frac{\delta^4}{12} + \frac{\delta^6}{90} - \cdots\right)y_i + h^2 f_i y_i = h^2 F_i \tag{6.79}$$

where h is increment in x. Multiplying both sides by $(1 + \delta^2/12)$, we get

$$\left(\delta^2 + \frac{\delta^6}{240} - \frac{13\delta^8}{15,120} + \cdots\right)y_i + h^2 f_i y_i + \frac{h^2 \delta^2}{12}(f_i y_i)$$

$$\tag{6.80}$$

$$= h^2 F_i + \frac{h^2 \delta^2}{12} F_i$$

Expanding terms involving δ^2 in terms of their arguments, we get

$$y_{i+1} - 2y_i + y_{i-1} + h^2 f_i y_i + \frac{h}{12}(f_{i+1}y_{i+1} - 2f_i y_i + f_{i-1}y_{i-1})$$

$$= h^2 F_i + \frac{h^2}{12}(F_{i+1} - 2F_i + F_{i-1}) - \left(\frac{\delta^6}{240} - \frac{13\delta^8}{15,120} + \cdots\right)y_i$$

$$\tag{6.81}$$

Neglecting the last expression in the above equation will result in an error of $0(h^6)$. Solving the above for y_{i+1}, we get

$$y_{i+1} = \frac{1}{1 + (h^2/12)f_{i+1}} \left[-\left(1 + \frac{h^2}{12}f_{i-1}\right)y_{i-1} + \left(2 - \frac{5h^2}{6}f_i\right)y_i \right.$$
$$\left. + \frac{h^2}{12}(F_{i-1} + 10F_i + F_{i+1}) \right] \tag{6.82}$$

which is Noumerov's recursive expression for the solution of Eq. (6.78).

Example. As an example of applying Noumerov's method, consider the problem of solving the differential equation

$$\frac{d^2x}{dt^2} + x = 0 \tag{6.83}$$

with the initial conditions $t = 0$, $x = 1.0$, and $dy/dx = 0$. Assume that the first few points of the solution are evaluated by Taylor's method of starting the solution, as follows:

$$
\begin{array}{lll}
t = 0 & x = 1.0 & \\
t = 0.05 & x = 0.99875 & \\
t = 0.10 & x = 0.99500 & \\
t = 0.15 & x = 0.98877 & (6.84)\\
t = 0.20 & x = 0.98007 & \\
t = 0.25 & x = 0.96891 &
\end{array}
$$

Comparing Eqs. (6.83) and (6.78), we get

$$f = 1 \quad \text{and} \quad F = 0 \tag{6.85}$$

Using these relations in Eq. (6.82), we get

$$x_{i+1} = \frac{1}{1 + (h^2/12)} \left[\left(1 + \frac{h^2}{12}\right)x_{i-1} - \left(\frac{5h^2}{6} - 2\right)x_i \right] \tag{6.86}$$

which is the recursion equation obtained from Noumerov's method. Having two values of x, that is, x_0 and x_1 corresponding to t_0 and t_1, the value of x_2 can be obtained from the above expression.

Now again consider Eq. (6.83), but this time let us substitute the central difference representation for the second derivative term as given in Table 5.4 of Chapter 5

$$\frac{x_{i+1} - 2x_i + x_{i-1}}{h^2} + x_i = 0 \tag{6.87}$$

From this we get

$$x_{i+1} = x_i(2 - h^2) - x_{i-1} \tag{6.88}$$

Note that the error in the above expression, from Table 5.4, is the order of h^2 while the corresponding error Eq. (6.86) is in the order of h^6. The recursion Eq. (6.88) also gives the value of x in terms of two previous values. Solutions given by Eqs. (6.86) and (6.88) are evaluated for Δt values of 0.1 and 0.20 and are given in Tables 6-6 and 6-7. The last columns of these tables give the values of the solution obtained from

$$x = \cos t \tag{6.89}$$

which is the analytical solution of the problem. In Tables 6-6 and 6-7 (E-2) represents 10^{-2}. From these tables it is seen that Noumerov's method of Eq. (6.86) is closer to the analytical solution for both $h = 0.1$ and 0.2. And, in addition, the central difference solution of Eq. (6.88) improves as the time step is reduced from 0.2 to 0.1. For example, the central difference solutions from Tables 6-6 and 6-7, for $t = 1$, are

$$h = 0.2 \quad t = 1.0 \quad x = 0.539189$$
$$h = 0.1 \quad t = 1.0 \quad x = 0.540302$$
$$\text{Analytical solution} \quad t = 1.0 \quad x = 0.540302$$

Thus the solution approaches the analytical solution as h assumes smaller values.

Table 6-6.   Solution of Eq. (6.83) for Time Step $\Delta t = h = 0.10$

Time t	x (Eq. (6.86))	x (Eq. (6.88))	x (Eq. (6.89))
0	1	1	1
.1	.995	.995	.995004
.2	.980058	.98005	.980067
.3	.955324	.955299	.955336
.4	.921045	.920996	.921061
.5	.877563	.877483	.877583
.6	.825312	.825194	.825336
.7	.764815	.764654	.764842
.8	.696677	.696467	.696707
.0	.621577	.621316	.62161
1.	.540267	.539951	.540302
1.1	.453559	.453187	.453596
1.2	.362319	.361891	.362358
1.3	.267459	.266976	.267499
1.4	.169926	. .169391	.169967
1.5	7.06955 E-2	7.01128 E-2	7.07372 E-2
1.6	−2.92413 E-2	−2.98669 E-2	−2.91995 E-2
1.7	−.128886	−.129548	−.128844
1.8	−.227243	−.227934	−.227202
1.9	−.323329	−.32404	−.32329
2.	−.416185	−.416906	−.416147

Table 6-7.   Solution of Eq. (6.83) for Time Step $\Delta t = h = 0.20$

Time t	x (Eq. (6.86))	x (Eq. (6.88))	x (Eq. (6.89))
0	1	1	1
.2	.980080	.980070	.980067
.4	.921067	.920937	.921061
.6	.825345	.824967	.825336
.8	.696718	.695998	.696707
1.	.540315	.539189	.540302
1.2	.362371	.360813	.362358
1.4	.169980	.168004	.169967
1.6	−2.91869 E-2	−3.15255 E-2	−2.91995 E-2
1.8	−.227190	−.229794	−.227202
2.	−.416137	−.418870	−.416147

The above solution can easily be programmed for a digital computer. The following is the program of the solution written in BASIC programming language. See [3].

```
5    DIM X(500),Y(500),Z(500)
10   READ H,X(0),X(1)
20   LET A=(((5/6)*H*H)-2)/(1+(1/12)*H*H)
30   LET B=2-H*H
40   LET Y(0)=X(0)
50   LET Z(0)=X(0)
60   LET Y(1)=X(1)
65   LET Z(1)=COS(H)
68   FOR I=1 TO 20
70   LET X(I+1)= -X(I-1)-A*X(I)
80   LET Y(I+1)=B*Y(I)-Y(I-1)
90   LET Z(I+1)=COS(H*(I+1))
100  NEXT I
110  PRINT "FOR H="H
120  PRINT
130  PRINT "TIME","NOUMEROV","CENTRAL DIFF","ANALYTIC SOL"
140  FOR I=0 TO 20
150  PRINT H*I,X(I),Y(I),Z(I)
160  NEXT I
170  DATA .1,1.,.99500
180  END
```

In this program X is the solution obtained from Eq. (6.86) by Noumerov's method, Y is the solution obtained from Eq. (6.88) by the central difference method, and Z is the analytical solution of Eq. (6.89). Statement 5 reserves memory storage for these variables. Statement 10 reads the data of initial values. Statement 20 computes the coefficient of x_i in Eq. (6.86). Statement 30 computes the coefficient of x_i in Eq. (6.88). Statements 40 through 65 set the initial values of central difference $Y(0)$, $Y(1)$, and analytical solution $Z(0)$ and $Z(1)$. Statements 68 through 100 compute the Noumerov solution given by Eq. (6.86), the central difference solution given by Eq. (6.88) and the analytical solution given by Eq. (6.89). The rest of the statements print out the solutions. The computer output corresponding to this run is given in Table 6-6.

Runge, Heun, and Kutta Integration Formulas

These integration formulas are widely used in the solution of differential equations with digital computers. Their primary advantage lies in the fact that no special method is needed for calculating starting values as was the case with some of the methods presented. For the complete derivation of the formulas, the reader is referred to [5]. In this section we will give the general method of this derivation and will present several commonly used integration formulas.

Taylor series expansion of the dependent variable y_i in terms of the independent variable x_i around the point x_i, y_i can be written

$$y_{i+1} = y_i + \Delta x y_i^{(1)} + \frac{\Delta x^2}{2!} y_i^{(2)} + \frac{\Delta x^3}{3!} y_i^{(3)} + \cdots \qquad (6.90)$$

where $\Delta x = x_{i+1} - x_i$ and the superscripts in parentheses specify the order of derivatives. Thus, having the values of x_i, y_i and the derivatives $y_i^{(1)}$, $y_i^{(2)}$, $y_i^{(3)}$, ... the values of y_{i+1} can be evaluated using Eq. (6.90). The value of x_{i+1} is, of course,

$$x_{i+1} = x_i + h \qquad (6.91)$$

where h is the increment in x. For a given differential equation

$$y^{(1)} = \frac{dy}{dx} = f(x, y) \qquad (6.92)$$

the values of the derivatives can be obtained as follows:

$$y^{(1)} = f(x, y)$$

$$y^{(2)} = \frac{\partial f}{\partial x} + \frac{\partial f}{\partial y}\frac{dy}{dx} = f_x + f_y f \qquad (6.93)$$

$$y^{(3)} = f_{xx} + 2f_{xy}f + f_{yy}f^2 + (f_x + f_y f)f_y$$

In these expressions x and y subscripts represent partial derivatives with respect to x and y variables. Evaluating these derivatives at x_i, y_i and setting in Eq. (6.90), we get

$$y_{i+1} = y_i + f_i h + \frac{1}{2}\left(f_x + f_y f\right)_i h^2$$

$$+ \frac{1}{6}\left[f_{xx} + 2f_{xy}f + f_{yy}f^2 + (f_x + f_y f)f_y\right]_i h^3 + \cdots \qquad (6.94$$

where subscripts i refer to values at x_i, y_i. This formula represents the third-order approximation to the values of y as the fourth-order terms (terms containing h^4) are neglected.

Equation (6.94) is the basic equation in developing Runge's integration formulas. Note that using only the first two terms on the right of the equal sign of Eq. (6.94) reduces this equation to Euler's method. In this case, the error will be of order h^2 instead of h^4 of Runge's formula.

Using the definitions

$$\Delta'y = f(x, y) \Delta x = fh$$
$$\Delta''y = f(x + mh, y_0 + m \Delta'y) h \tag{6.95}$$
$$\Delta'''y = f[x + \lambda h, y_0 + \rho \Delta''y + (\lambda - \rho) \Delta'y] h$$

where m, λ, and ρ are constants, it is possible to expand the above functions in Taylor series and sum up the resulting equations in the form

$$\Delta y = y_{i+1} - y_i$$
$$= a \Delta'y + b \Delta''y + c \Delta'''y \tag{6.96}$$

where a, b, and c are again constants. Equating the terms of Eq. (6.96) with those of Eq. (6.95), after a considerable amount of algebra, relations between the constants m, λ, ρ, a, b, and c can be obtained. Runge's formulas are obtained by making certain assumptions in using these relations. For the complete derivation the reader is referred to [5]. Here, however, we will give several commonly used integration formulas obtained by the above procedure.

1. *Heun's formula:*

$$\Delta y_i = y_{i+1} - y_i$$
$$= \tfrac{1}{4}(\Delta'y + 3\Delta'''y)_i \tag{6.97}$$

where

$$\Delta'y = f_i(x, y) h$$

$$\Delta''y = f_i\left(x + \tfrac{1}{3}h, y + \tfrac{1}{3}\Delta'y\right) h$$

$$\Delta'''y = f_i\left(x + \tfrac{2}{3}h, y + \tfrac{2}{3}\Delta''y\right) h$$

2. *Kutta's third-order rule:*

$$\Delta y_i = \frac{1}{6}(\Delta'y + 4\Delta''y + \Delta'''y)_i \qquad\qquad (6.98)$$

where

$$\Delta'y = f_i(x, y)\, h$$

$$\Delta''y = f_i\!\left(x + \frac{1}{2}h,\ y + \frac{1}{2}\Delta'y\right)h$$

$$\Delta'''y = f_i(x + h,\ y + 2\Delta''y - \Delta'y)\, h$$

Note that this formula is analogous to Simpson's one-third rule.

3. *Kutta's fourth-order approximation (Kutta-Simpson one-third rule):*

$$\Delta y_i = \frac{1}{6}(\Delta'y + 2\Delta''y + 2\Delta'''y + \Delta^{iv}y)_i$$

where

$$\Delta'y = f_i(x, y)\, h$$

$$\Delta''y = f_i\!\left(x + \frac{1}{2}h,\ y + \frac{1}{2}\Delta'y\right)h$$

$$\Delta'''y = f_i\!\left(x + \frac{1}{2}h,\ y + \frac{1}{2}\Delta''y\right)h \qquad (6.99)$$

$$\Delta^{iv}y = f_i(x + h,\ y + \Delta'''y)\, h$$

4. *Kutta-Simpson three-eighths rule:*

$$\Delta y_i = \frac{1}{8}(\Delta'y + 3\Delta''y + 3\Delta'''y + \Delta^{iv}y)$$

where

$$\Delta'y = f_i(x, y)\,h$$

$$\Delta''y = f_i\left(x + \frac{1}{3}h,\ y + \frac{1}{3}\Delta'y\right)h$$

$$\Delta'''y = f\left(x + \frac{2}{3}h,\ y + \Delta''y - \frac{1}{3}\Delta'y\right)h$$

(6.100)

$$\Delta^{iv}y = f(x + h,\ y + \Delta'''y - \Delta''y + \Delta'y)\,h$$

The error in using Eqs. (6.99) and (6.100) is of the order h^5 as opposed to order h^4 of Eqs. (6.97) and (6.98). In the derivation of the fourth-order equations an additional term containing h^4 was utilized in Eq. (6.94).

As mentioned previously, in using Runge's formulas no special method is needed in obtaining the initial values as was the case of predictor-corrector methods.   In many applications the requirement of using a special starting method is considered a disadvantage. For this reason and because of its accuracy, Runge's method is widely used in digital computer solution of differential equations.

Example. As an example of Runge's method of solution, consider the differential equation

$$f(x, y) = \frac{dy}{dx} = e^{-3x} - y, \quad y(0) = 0.5$$

(6.101)

The analytic solution of this equation is

$$y = e^{-x} - 0.5e^{-3x}$$

(6.102)

The program can be formulated for numerical solution by using, for example, Kutta's third-order rule of Eq. (6.98). This can be done by assuming a value for $\Delta x = h$ and systematically evaluating $\Delta'y$, $\Delta''y$, $\Delta'''y$ of Eqs. (6.98) at $x_0 = 0$, $y_0 = 0.5$. The resulting values are then used in the first of Eqs. (6.98) to obtain

$$y_1 = y_0 + \Delta y_0$$

(6.103)

Similarly, for y_2, y_3, and so on. For comparison, solutions were also evaluated using only the first two terms of Eq. (6.94)—Euler's method,

$$y_{i+1} = y_i + f_i h \qquad (6.104$$

The results of these evaluations are given in Table 6-8 for h values of 0.01, 0.05, and 0.1. From this table it is seen that Kutta's formula (6.98) gives answers correct to four significant figures with an $h = 0.1$ while Euler's method can be considered acceptable at $h = 0.01$.

In practice, the value of the increment h to be used is arrived at by experimenting with the solution on the computer. In the case of the above problem, the problem is run with $h = 0.1$ and $h = 0.05$; if the answers do not change appreciably in the region of interest, the higher value of $h(0.1)$ is chosen. The mere fact that the solution changes when the time or space increment is changed points to the fact that inherent errors existing in the finite difference representation of the problem are magnified by the value of the increment. For example, consider the values of the solution by Euler's method of Table 6-8 at $t = 1.0$. It is noted that the solutions vary as h is decreased from 0.1 to 0.05 to 0.01 If this were the only method being used one might use $h = 0.005$ to see if the compounding of errors degrades the solution before deciding on the final value of h. However, in decreasing the values of h the reader should be aware of the problems of the stability of solution which might arise.

Computer Program for the Evaluation of the Above Solution. The following computer program in a version of FORTRAN programming language [6] was written to implement the above solution. In the program, Y2 refers to solution obtained by Eq. (6.98), Y3 by Eq (6.104), and Y4 by Eq. (6.102). Statement 110 defines the function of Eq. (6.101). Statements 120 through 150 are constants. Statements 170 through 220 are mechanization of Eqs. (6.98), while statement (230) represents Eq. (6.104). The analytical solution of Eq. (6.102) is given by statement 250. The results of the program are given in Table 6-8.

```
100   DIMENSION Y2(250),Y3(250),Y4(250),X(250)
110   FXY (X,Y)=EXP(-3.*X)-Y
120   Y2(0)=.5
130   Y3(0)=.5
```

Table 6-8.   Kutta's Third-order Rule in Solving Eq. (6.101)

Time t	Kutta equation (6.98)	Euler equation (6.104)	Analytic solution (6.102)	Increment $\Delta x = h$
.00	.50	.50	.50	
.10	.5344	.5358	.5344	
.20	.5443	.5466	.5443	
.30	.5375	.5402	.5375	
.40	.5197	.5226	.5197	
.50	.4950	.4978	.4950	
.60	.4662	.4689	.4662	
.70	.4354	.4379	.4354	
.80	.4040	.4062	.4040	
.90	.3730	.3750	.3730	$h = 0.01$
1.00	.3430	.3447	.3430	
1.10	.3144	.3159	.3144	
1.20	.2875	.2888	.2875	
1.30	.2624	.2635	.2624	
1.40	.2391	.2400	.2391	
1.50	.2176	.2183	.2176	
1.60	.1978	.1983	.1978	
1.70	.1796	.1801	.1796	
1.80	.1630	.1634	.1630	
1.90	.1479	.1481	.1479	
2.00	.1341	.1342	.1341	
.00	.50	.50	.50	
.50	.4950	.5097	.495	
1.00	.3430	.3521	.343	
1.50	.2176	.2213	.2176	
2.00	.1341	.1348	.1341	$h = 0.05$
2.50	.0818	.0812	.0818	
3.00	.0497	.0488	.0497	
3.50	.0302	.0292	.0302	
4.00	.0183	.0175	.0183	
4.50	.0111	.0105	.0111	
5.00	.0067	.0063	.0067	
0.00	.50	.50	.50	
1.00	.3430	.3621	.3430	
2.00	.1341	.1356	.1341	$h = 0.1$
3.00	.0497	.0477	.0497	
4.00	.0183	.0167	.0183	
5.00	.0067	.0058	.0067	

```
140  Y4(0)=.5
150  H=.05
160 ┌DO10I=0,199
170 │AI=I
180 │X(I)=AI*H
190 │D1Y=FXY(X(I),Y2(I))*H
200 │D2Y=FXY(X(I)+H/2.,Y2(I)+D1Y/2.)*H
210 │D3Y=FXY(X(I)+H,Y2(I)+2.*D2Y-D1Y)*H
220 │Y2(I+1)=Y2(I)+(D1Y+4.*D2Y+D3Y)/6.
230 │Y3(I+1)=Y3(I)+FXY(X(I),Y3(I))*H
240 │X(I+1)=X(I)+H
250 │Y4(I+1)=EXP(-X(I+1))-EXP(-3.*X(I+1))/2.
260 └10
270  PRINT,"    TIME        Y2        Y3        Y4"
280  PRINT
290 ┌DO 20 I=0,200,10
300 │PRINT,X(I),Y2(I),Y3(I),Y4(I)
310 └20
320  STOP
330  END
```

Runge-Kutta Formulas for a Set of Simultaneous Equations

As discussed in the beginning of this chapter, higher-order differential equations can be reduced to sets of simultaneous equations and these equations solved by various methods. One method in common use is the Runge-Kutta method, which is in turn an extension of the given formulas for the solution of first order equations. Consider the simultaneous equations

$$\frac{dy}{dx} = f(x, y, z)$$

$$\frac{dz}{dx} = g(x, y, z)$$

(6.105)

For example, Kutta's fourth-order approximation of Eq. (6.99) in this case becomes

$$\Delta y_i = \frac{1}{6}(\Delta'y + 2\Delta''y + 2\Delta'''y + \Delta^{iv}y)$$

$$\Delta z_i = \frac{1}{6}(\Delta'z + 2\Delta''z + 2\Delta'''z + \Delta^{iv}z)$$

(6.106)

where (for $\Delta x = h$)

$$\Delta' y = f_i(x, y, z) h$$

$$\Delta' z = g_i(x, y, z) h$$

$$\Delta'' y = f_i\left(x + \frac{1}{2} h, y + \frac{1}{2} \Delta' y, z + \frac{1}{2} \Delta' z\right) h$$

$$\Delta'' z = g_i\left(x + \frac{1}{2} h, y + \frac{1}{2} \Delta' y, z + \frac{1}{2} \Delta' z\right) h$$

$$\Delta''' y = f_i\left(x + \frac{1}{2} h, y + \frac{1}{2} \Delta'' y, z + \frac{1}{2} \Delta'' z\right) h \qquad (6.107)$$

$$\Delta''' z = g_i\left(x + \frac{1}{2} h, y + \frac{1}{2} \Delta'' y, z + \frac{1}{2} \Delta'' z\right) h$$

$$\Delta^{iv} y = f_i(x + h, y + \Delta''' y, z + \Delta''' z) h$$

$$\Delta^{iv} z = g_i(x + h, y + \Delta''' y, z + \Delta''' z) h$$

If these increments in y and z are computed in the order given, only previously computed increments will be needed at each step in computation.

Solution by Direct Substitution

In a great number of problems it is possible to obtain solutions by directly substituting difference equations in place of differentials of a given differential equation. This was done in the case of the example problem in Eq. (6.87). In many closed-loop servo problems this method of direct substitution is the most practical way of solving the problem. In applying the method, care should be exercised to choose a time step which will give satisfactory solution. In engineering applications, this time step is chosen by experiments. That is, solution is obtained for a certain value of the time step and then this value of the time step is halved to see if the solution changes. If the solution remains practically the same, the original value of the time step is doubled and a solution obtained. The idea behind this is to obtain the largest practical value of the time step without detracting from the accuracy of the solution.

To illustrate the method of direct substitution, let us consider the following closed-loop navigation problem.

Example. In this example we will illustrate the digital computer simulation of the proportional navigation which is often employed in missile guidance [7]. To derive the appropriate differential equations of this navigation scheme, consider the two-dimensional geometry of Fig. 6-1. In this figure a missile centered coordinate system is used. The missile-to-target line-of-sight vector is designated by $\bar{R}$ and the closing

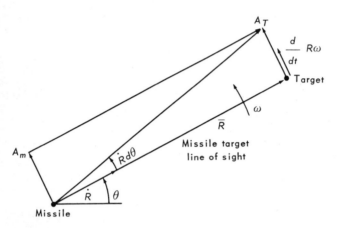

figure 6-1. *Two-dimensional diagram of proportional navigation.*

rate along this vector as $\dot{R}$. The navigation scheme is based on the following relation

$$A_M = A_T = A_R \tag{6.108}$$

which states that the acceleration of the missile perpendicular to $\bar{R}$, A_M, should be equal to the acceleration of the target perpendicular to $\bar{R}$, A_T, and equal to the rotational acceleration of line-of-sight $\bar{R}$, A_R. From this figure it is noted that this will keep the line-of-sight $\bar{R}$ parallel and since $\dot{R}$ is the rate of decrease of the vector $\bar{R}$, an intercept is assured. The line-of-sight acceleration A_R will consist of two parts, the acceleration perpendicular to $\bar{R}$ which will equal the rate of change of tangential velocity $R\omega$ where ω is the angular rate, and the vectorial change of $\dot{R}(=\dot{R}\,d\theta)$. This will result in

$$A_R = \frac{d}{dt}(R\omega) + \dot{R}\frac{d\theta}{dt} \tag{6.109}$$

Differentiating this expression and using the relation

$$\omega = \frac{d\theta}{dt} \tag{6.110}$$

where θ is the angle of $\bar{R}$ with respect to some reference

$$A_R = R\dot{\omega} + 2\omega\dot{R} \tag{6.111}$$

Setting this value in Eq. (6.108) results in

$$A_T - A_M = R\dot{\omega} + 2\omega\dot{R} \tag{6.112}$$

Using differential operators D and D^2, Eq. (6.112) becomes

$$A_T - A_M = R\,D^2\theta + 2\dot{R}\,D\theta \tag{6.113}$$

or

$$\theta = \frac{A_T - A_M}{R\,D^2 + 2\dot{R}\,D} \tag{6.114}$$

This equation can be written in servo notation of [8]. The servo loop of Eq. (6.114) can be obtained by assuming that the missile acceleration A_M is proportional to the angular rate of θ or $D\theta$. In addition, let us assume the current value of range R as given by the equation

$$R = R_0 + t\dot{R} \tag{6.115}$$

where R_0 is the initial value of R at $t = 0$ and t is the time. Using this relation in Eq. (6.114) we can draw the servo loop shown in Fig. 6-2. In this figure at point a we have the numerator of Eq. (6.114) and at point b we have the value of θ. Assuming that the commanded missile acceleration is proportional to $D\theta$, we have

$$A_M = k\,D\theta \tag{6.116}$$

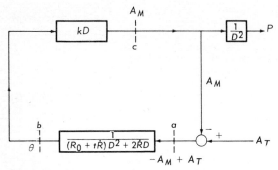

figure 6-2. *Servo loop of proportional navigation problem.*

where k is the proportionality constant. Using Eq. (6.116) we get to point c of the loop of Fig. 6-2. From this point we get back to a and the loop is closed. Note that integrating acceleration A_M twice will give the position of the missile, P of Fig. 6-2.

As a numerical example of simulating the loop of Fig. 6-2, consider the following values and Fig. 6-3. A missile with a velocity of 1,500 fps

Missile ● $V_2 = 1,500$ ft/sec $V_1 = 500$ ft/sec ● Target

|←————————20,000 ft————————→|

figure 6-3. *Numerical values of the proportional navigation problem.*

is fired at an approaching target with a velocity of 500 fps at a range of 20,000 ft. At this instant the target maneuvers with a vertical acceleration of 32 fps, keeping its horizontal velocity constant. Assuming that the missile horizontal velocity remains constant, what are the missile vertical accelerations, using proportional navigation? To write the digital simulation of the problem, we use the central difference methods of Chap. 5. Let us denote

$$A_T - A_M = U \qquad (6.117)$$

Equation (6.114) can be written, using central differences for the derivatives as

$$\left(R_0 + t\dot{R}\right)_i \frac{\theta_{i+1} - 2\theta_i + \theta_{i-1}}{(\Delta t)^2} + 2\dot{R} \frac{\theta_{i+1} - \theta_{i-1}}{2\Delta t} = U_i \qquad (6.118)$$

This equation can be solved for θ_{i+1} in terms of θ_i and θ_{i-1}. Having the value of θ_{i+1}, point b of the loop of Fig. 6-2, the value of missile acceleration A_M can be obtained from (6.116):

$$\left(A_M\right)_{i+1} = k\,\frac{\theta_{i+1} - \theta_i}{\Delta t} \tag{6.119}$$

The position of the missile P can be obtained from

$$D^2 P = A_M \tag{6.120}$$

or, in finite difference form,

$$\frac{P_{i+1} - 2P_i + P_{i-1}}{\Delta t^2} = \left(A_M\right)_i \tag{6.121}$$

This equation can be solved for P_{i+1} in terms of P_i and P_{i-1}. Thus the loop of Fig. 6-2 can be started by setting initial values of θ, A_M, and P equal to zero and introducing target acceleration A_T. The positions of the missile and target with respect to the initial missile position can be obtained by using the following relation:

$$
\begin{aligned}
\text{Missile (horizontal)} &= V_2 t = 1{,}500t \\[4pt]
\text{Target (horizontal)} &= R_0 - V_1 t = 20{,}000 - 500t \\[4pt]
\text{Missile (vertical)} &= P \text{ (Eq. (6.121))} \\[4pt]
\text{Target (vertical)} &= \tfrac{1}{2} A_T t^2 = 16t^2
\end{aligned}
\tag{6.122}
$$

The following digital computer program has been written in BASIC programming language (see [3]) to accomplish the above procedure. In the program, the following notations are used:

L = angle θ in radians
A = missile acceleration, ft/sec^2

P = missile vertical position, ft

RO = R_0 initial value of $\overline{R}$, ft

$V1$ = target horizontal velocity, ft/sec

$V2$ = missile horizontal velocity, ft/sec

K = proportionality constant of Eq. (6.116) (ft/sec^2)/(rad/sec)

$A1$ = target vertical acceleration, ft/sec^2

$T2$ = time step size Δt, sec

$R1$ = $\dot{R}$ or closing rate = $-(V1 + V2)$

```
10    DIM L(500),A(500),P(500)
20    READ R0,V1,V2,K,A1,T2
25    LET R1=-(V1+V2)
30    PRINT "RZERO"RO,"TGT VEL"V1,"MSL VEL"V2
35    PRINT "K"K,"TGT ACC" A1,"DEL T"T2
36    PRINT
40    LET L(0)=0
41    LET L(1)=0
45    LET A(0)=0
46    LET A(1)=0
50    LET P(0)=0
51    LET P(1)=0
65    LET I1=RO/(-R1*T2)
70    FOR I=1 TO I1-2
75    LET T=I*T2
80    LET U=A1-A(I)
90    LET R2=R0+T*R1
100   LET L(I+1)=U*T2*T2+L(I)*(2*R2)+L(I-1)*(R1*T2-R2)
110   LET L(I+1)=L(I+1)/(R2+R1*T2)
120   LET A(I+1)=((L(I+1)-L(I))/T2)*K
130   LET P(I+1)=T2*T2*A(I)+2*P(I)-P(I-1)
140   NEXT I
145   LET P(I1)=T2*T2*A(I1-1)+2*P(I1-1)-P(I1-2)
150   PRINT "TIME","TGT POS H","TGT POS V",
160   PRINT "MSL POS H","MSL POS V"
170   FOR J=0 TO I1 STEP 20
180   PRINT J*T2,RO-J*T2*V1,(A1/2)*(J*T2)↑2,J*T2*V2,P(J)
190   NEXT J
200   LET J=I1
205   PRINT
210   PRINT J*T2,RO-J*T2*V1,(A1/2)*(J*T2)↑2,J*T2*V2,P(J)
220   PRINT
230   PRINT "TIME","ANGLE","MSL VER ACC"
235   FOR J=0 TO I1 STEP 20
```

```
240   PRINT J*T2,57.3*L(J),A(J)
250   LET J=I1
255   PRINT
260   PRINT J*T2,57.3*L(J),A(J)
270   DATA 20000,500,1500,20000,32,.025
280   END
```

Statement 10 of the program reserves storage locations for L, A, and P. Statement 20 reads the data as given in Statement 270. These values are

RO = 20,000 ft
$V1$ = 500 ft/sec
$V2$ = 1,500 ft/sec
K = 20,000 (ft/sec^2)/(rad/sec)
$A1$ = 32 ft/sec^2
$T2$ = 0.025 sec

Statement 25 sets the value of rate of change of R equal to the sum of $V1$ and $V2$ with a negative sign (since R decreases as the missile and target approach each other). It is assumed here that the angular change of vector $\bar{R}$ is small and $\bar{R}$ remains practically horizontal during the flight. Statements 30 through 36 are print statements which write out the input data. Statements 40 through 51 set initial conditions equal to zero. Statement 65 computes the upper bound of time index I. This value will be equal to the initial range divided by distance moved at each time step. Statements 70 through 140 are the main loop of the program which compute $L(I+1) = \theta_{i+1}$ from Eq. (6.118), $A(I + 1) = (A_M)_{i+1}$ from Eq. (6.119), and $P(I + 1) = P_{i+1}$ from Eq. (6.121). Note that the last position of the missile $P(I1)$ is computed in statement 145, since at the last point the missile-to-target angle L according to statement 110 will go to infinity as the denominator of L goes to zero. Statements 150 through 190 print time, horizontal target position (TGT POS H), vertical target position (TGT POS V) and horizontal and vertical missile positions (MSL POS H and MSL POS V), respectively. Note that in print statement 180 Eqs. (6.122) are used. Since statement 170 is indexed every 20 time steps there may be cases where the values at the last computed time step will not be printed. For this reason statement 210 prints the last computed values.

Statements 230 through 260 print out time, angle $L = \theta$ (in degrees), and missile vertical acceleration A (MSL VER ACC).

The output listing of the computer run is given in Table 6-9. (This output is obtained by combining the outputs of statements 170 through 260.)   A graph of the horizontal distance and vertical distance with time ticks is shown in Fig. 6-4.

Nonlinearities and Random Inputs

In digital computer programs nonlinearities describing the limits of variables can be entered through logical instructions. For example, if it is desired to limit the missile acceleration A_M of Fig. 6-2 at point c to a predetermined value of $\pm k_1$, this can be entered as a logical operation at point c and correspondingly in the computer program. The following flow chart describes this.

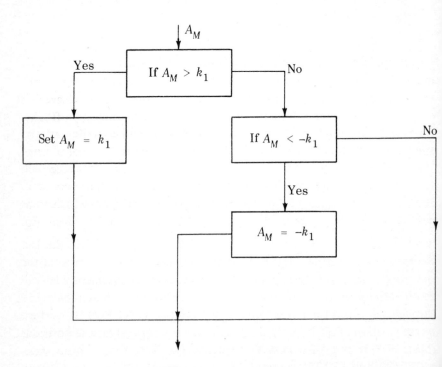

Random variables can be introduced equally well into digital computer programs. Almost all of the digital computer installations have built-in routines for the generation of random numbers y, with uniform

Table 6-9. Solution of Proportional Navigation Problem

TIME	TGT POS H	TGT POS V	MSL POS H	MSL POS V	ANGLE	MSL VER ACC
0	20000	0	0	0	0	0
0.5	19750	4.	750.	.531952	9.85574 E-3	13.0501
1.0	19500	16.	1500.	4.23719	3.61716 E-2	22.6065
1.5	19250	36.	2250.	13.5298	7.37154 E-2	29.0553
2.0	19000	64.	3000.	30.0407	.118751	33.3065
2.5	18750	100.	3750.	54.8469	.168664	36.0361
3.0	18500	144.	4500.	88.6408	.221664	37.7372
3.5	18250	196.	5250.	131.855	.276557	38.7616
4.0	18000	256.	6000.	184.751	.332569	39.3546
4.5	17750	324.	6750.	247.479	.389215	39.6823
5.0	17500	400.	7500.	320.125	.446202	39.8539
5.5	17250	484.	8250.	402.733	.503364	39.9379
6.0	17000	576.	9000.	495.324	.560607	39.976
6.5	16750	676.	9750.	597.908	.617886	39.9915
7.0	16500	784.	10500.	710.489	.675179	39.9972
7.5	16250	900.	11250.	833.069	.732477	39.9988
8.0	16000	1024.	12000.	965.649	.789776	39.9995
8.5	15750	1156.	12750.	1108.23	.847075	39.9996
9.0	15500	1296.	13500.	1260.81	.904374	39.9996
9.5	15250	1444.	14250.	1423.38	.961674	39.9998
10.0	15000	1600.	15000.	1595.96	0	0
10.0	15000	1600	15000	1595.96	0	0

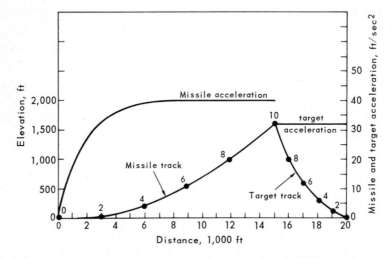

figure 6-4. *Graphical representation of solution given in Table 6-9.*

distribution, between zero and one.   For this distribution mean and standard deviations are given as

$$\text{Mean} = \mu(y) = \frac{1}{2}$$

(6.123)

$$\text{Standard deviation} = \sigma^2(y) = \frac{1}{12}$$

Since many of the errors in engineering systems can be represented by normally distributed white noise,* it is desirable to obtain random numbers with this distribution.   Using numbers y from a uniform distribution $(0, 1)$, it is possible to obtain numbers v from a normal distribution $N(\mu, \sigma)$, where μ is the mean and σ is the standard derivation.   To accomplish this the following procedures are used:

1.   Approximate   method—From   the   central   limit   theorem   [9] using Eqs. (6.123) we can obtain the following relations

$$v = \left[\lim_{k \to \infty} \frac{\displaystyle\sum_{i=1}^{k} y_i - \frac{k}{2}}{\sqrt{\dfrac{k}{12}}} \right] \sigma + \mu$$

(6.124)

*The term white noise is used when errors are independent of each other and are not time correlated.

For most applications $k \geq 6$ produces a distribution of v closely approximating a normal distribution $N(\mu, \sigma)$. From Eq. (6.124) $k = 12$ is a particularly desirable choice, since the division by the radical is eliminated and Eq. (6.124) reduces to the form

$$v = \left(\sum_{i=1}^{12} y_i - 6 \right) \sigma + \mu \tag{6.125}$$

In this expression, y_i, as before, represents a random number from a uniform distribution $(0, 1)$.

2. Exact method—Using statistical analysis (see [10]), it is possible to obtain two random numbers from a normal distribution $N(0, 1)$ by using two random numbers from a uniform distribution $(0, 1)$. Assuming that the random numbers are y_1 and y_2, we have

$$z_1 = \sqrt{-2 \ln y_1} \, \cos(2\pi y_2)$$
$$z_2 = \sqrt{-2 \ln y_1} \, \sin(2\pi y_2) \tag{6.126}$$

where z_1 and z_2 are two random numbers from a normal distribution $N(0, 1)$. To transform these to two random numbers from a normal distribution with mean value of μ and standard derivation σ, $N(\mu, \sigma)$, we write

$$v_1 = z_1 \sigma + \mu$$
$$v_2 = z_2 \sigma + \mu \tag{6.127}$$

where v_1 and v_2 are numbers from the normal distribution $N(\mu, \sigma)$.

Method 2 is more accurate but it uses more computer time as it requires the evaluation of logarithmic and trigonometric functions. The first method gives sufficiently accurate results and is commonly used.

In the previous example of Fig. 6-2, white noise of $N(\mu, \sigma)$ can be introduced, for example, on the values of missile acceleration A_M by computing v values from the above equations and adding to the values of acceleration at each time step. The bias errors (if they exist) are treated as constants and are directly added to the values of the variables.

Aside from the problems discussed above, in many computer applications it is desired to obtain estimates of the *mean* $\bar{x}$ and the *standard deviation* s of random parameters. These values are computed using the equations

$$\bar{x} = \frac{1}{M} \sum_{i=1}^{M} x_i$$

$$s = \sqrt{\sum_{i=1}^{M} \frac{(x_i - \bar{x})^2}{M-1}} \tag{6.128}$$

$$= \sqrt{\frac{\sum_{i=1}^{M} x_i^2}{M-1} - \bar{x}^2}$$

where x_i is the parameter or the variable and M is the number of samples available. The last equation for s is more suitable for digital computer calculations since the value of $\bar{x}$ is not needed for keeping the running sums of x_i^2. The need for the computation of the mean and standard deviation primarily comes in reduction of data and in the case of Monte Carlo programs. These are programs with random inputs which are run many times to obtain statistical distribution of the parameters of the solution. For example, introducing noise with a normal distribution, as discussed above, at point c of Fig. 6-2 will result in parameters such as P and θ as a function of time with noise superposed. If the statistical behavior of these parameters at certain instances of time is required, the program should be run many times (say, about 100)* and the mean and sigma values calculated at these required times.

PROBLEMS

1. Obtain Taylor series expansion of the solution of the equation

*The number of times a Monte Carlo program is run (or equivalently the number of data points) depends on the desired accuracy and confidence intervals of statistical parameters (see [9]).

$$\frac{dx}{dt} = 6 - 3x$$

about the initial condition $x = 0$ at $t = 0$. From this series, obtain the value of x at $t = 0.2$. Compare with analytical solution.

2. Obtain Taylor series solution of the simultaneous equations

$$\frac{dz}{dx} = x + y$$

$$\frac{dy}{dx} = x + z$$

about the initial conditions $x = 0$, $z = 1$, $y = 1$.

3. Obtain the Taylor series expansion of the equation

$$\frac{d^2 y}{dx^2} = -3x^2 + 2y$$

about the initial conditions $x = 1$, $y = 1$, $dy/dx = 0$. From this series, obtain the values of y for $x = 1.1$, 1.2, and 1.3.

4. Obtain the solution of the differential equation

$$-\frac{dy}{dx} = \frac{4 - x^2 + y^2}{4y}$$

with the initial conditions $y = 1$ for $x = 2$, for $x = 2.1$, 2.2, and 2.3, using Euler's method.

5. Obtain the solution of the differential equation

$$\frac{dy}{dx} = \frac{1 - x - 2y}{y + 3}$$

with the initial conditions $x = 3$, $y = 1$, for $x = 3.05$ and 3.10, using Euler's method.

6. Obtain the solution of the differential equation

$$\frac{dy}{dx} = \frac{e^y + 4x}{y^2}$$

with the initial conditions $y = 1$ when $x = 1$, for $x = 1.1$ and 1.2.

7. Using Milne's method, formulate the numerical solution of the differential equation

$$\frac{dy}{dx} = x^3 + 3y$$

assuming that y_0, y_1, y_2, and y_3 corresponding to x_0, x_1, x_2, and x_3 are available.

8. Using Milne's method, formulate the numerical solution of the differential equation

$$\frac{dy}{dt} = y^2 + t^3$$

assuming that y_0, y_1, y_2, and y_3 corresponding to t_0, t_1, t_2, and t_3 are given.

9. Derive Eq. (6.51) from Eq. (6.50).

10. Formulate the solution of the differential equation

$$\frac{dy}{dx} = x^3 + 3y$$

with the condition $x = 0$, $y = 1$, by the Fox-Euler method.

11. Point out the difficulty encountered in using the Fox-Euler method in the solution of the nonlinear differential equation

$$\frac{dy}{dx} = x^2 + y^2$$

with the initial conditions $x = x_0$, $y = y_0$.

12. Formulate the solution of the above problem by Euler's method of starting the solution.

13. Formulate the solution of the second-order differential equation

$$\frac{d^2 y}{dx^2} = 3\frac{dy}{dx} + x^2$$

with the initial conditions $x = 0$, $y = 1$ $dy/dx = 2$, by Milne's predictor-corrector method.

14. Formulate the solution of the above equation (Prob. 13) by direct substitution of central differences for the derivatives.

15. Formulate the solution of the differential equation

$$\frac{d^2y}{dx^2} = x^2 + 3y$$

with the initial conditions $x = 1$, $y = 1$, by Noumerov's method.

16. Include the term $\dfrac{\Delta x^4}{4!}\, y_i^{(4)}$ in Eq. (6.90) and obtain the corresponding equation to Eq. (6.94)

17. Evaluate the solutions given in Table 6-8 for $h = 0.01$ up to time of 0.1.

18. Evaluate the solutions given in Table 6-8 for $h = 0.1$ up to time of 1.0.

19. Formulate the solution of simultaneous equations of Prob. 2 by the Runge-Kutta method.

20. Formulate the solution of differential equation of Prob. 13 by the Runge-Kutta method.

21. Obtain the corresponding equations to Eqs. (6.106) and (6.107) using Kutta's third-order rule of Eq. (6.98).

22. Obtain the corresponding equations to Eqs. (6.106) and (6.107) using Kutta-Simpson three-eighths rule of Eqs. (6.100).

23. Formulate the solution of Prob. 15 by the Runge-Kutta method.

24. Write the proportional navigation problem of Fig. 6-2 in terms of velocities. In this case Eq. (6.108) becomes

$$V_M = V_T = V_R$$

where V_M and V_T are the velocities of missile and target perpendicular to $\bar{R}$. In this equation, $V_R = R\omega$ and

$$V_M = V_T - R\omega$$

Differentiating, we get

$$\frac{dV_M}{dt} = \frac{dV_T}{dt} - \dot{R}\omega - \dot{\omega}R$$

or

$$A_M = A_T - \dot{R}\omega - \dot{\omega}R$$

Carry out the rest of the formulation from Eq. (6.108) using the above relation.

25. Draw the logic block diagram to represent the nonlinear function y as a function of x.

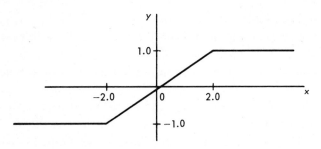

Prob. 25

26. Draw the logic block diagram to represent the nonlinear function y as a function of x

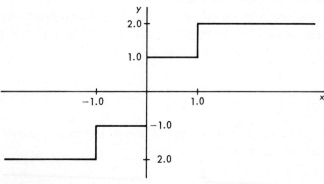

Prob. 26

27. Write the logic of entering normally distributed white noise with a $\mu = 0.005$ and $\sigma = 0.01$ ft/sec^2 at point c of Fig. 6-2.

REFERENCES

1. Boyce, William E., and Richard C. DiPrima: "Elementary Differential Equations," John Wiley & Sons, Inc., New York, 1969.
2. Milne, W. E., and R. R. Reynolds: Stability of a Numerical Solution of Differential Equations, *J. ACM*, vol. 6, p. 196, 1959.
3. BASIC LANGUAGE, reference manual, General Electric Company, May, 1966.
4. Chase, P. E.: Stability Properties of Predictor-Corrector Methods for Ordinary Differential Equations, *J. ACM*, vol. 9, p. 457, 1962.
5. Kunz, Kaiser S.: "Numerical Analysis," McGraw-Hill Book Company, New York, 1957.
6. "Time Sharing FORTRAN Reference Manual," General Electric Company, October, 1966.
7. Hovanessian, S. A., W. W. Maguire, and D. K. Richardson, Mathematical Models in the ASME Development of a Complex Missile System, *Aviation and Space Conference Proceedings,* June, 1968, p. 681.
8. D'Azzo, J. J., and C. H. Houpis: "Feedback Control System Analysis and Synthesis," McGraw-Hill Book Company, New York, 1960.
9. Hoel, Paul G.: "Introduction to Mathematical Statistics," pp. 107, 226, John Wiley & Sons, Inc., New York, 1958.
10. Box, G. E. P., and Mervin E. Muller: "A Note on the Generation of Random Normal Deviates," *Ann. Math. Stat.,* vol. 29, pp. 610-611, 1958.
11. Hildebrand, F. B.: "Introduction to Numerical Analysis," McGraw-Hill Book Company, New York, 1956.

7 BOUNDARY VALUE DIFFERENTIAL EQUATIONS AND STABILITY OF SOLUTION OF DIFFERENTIAL EQUATIONS

BOUNDARY VALUE PROBLEMS

From the theory of linear differential equations it is known that the general solution of an nth-order differential equation will contain n arbitrary constants. These constants are evaluated from the given conditions of the dependent variable and its derivatives at certain values of the independent variable. As discussed in the previous chapter, if all of these conditions are specified for the value of the independent variable x_0, where the solution is desired for $x > x_0$, then the differential equation is termed an initial value problem. If some of the conditions of the dependent variable are specified at the end point $x = L$, where the solution is desired for $x_0 \le x \le L$, then the problem is termed a *two-point* boundary value problem. In this chapter the numerical solution of the boundary value problems will be considered primarily for linear differential equations in order to compare results with analytical solutions. The methods described can equally well be applied to nonlinear differential equations.

The methods of the previous chapter, together with relations given in Chap. 5, can be applied to the solution of boundary value problems. We will show that the numerical solutions of the boundary value problems can be accomplished either by iterative methods or can be reduced to the solution of a set of n simultaneous linear equations in n unknowns. In this case the number of equations will depend on the step size (delta) of the independent variable. The methods of numerical solution of linear simultaneous equations discussed in Chap. 1 can readily be applied to evaluate the unknowns of the simultaneous equations.

Iterative Methods

Consider the differential equation

$$\frac{d^m y}{dx^m} = f(x, y, y', y'', \ldots, y^{m-1}) \tag{7.1}$$

where the superscripts refer to derivatives with respect to x. The above differential equation will require m auxiliary conditions for solution. Using finite-difference representations, Eq. (7.1) can be put in finite-difference form:

$$y_{i+1} = f(\ldots y_{i-3}, y_{i-2}, y_{i-1}, y_i, \Delta x) \tag{7.2}$$

where y_{i+1} is expressed in terms of its values at previous x positions and the increment Δx. Since in a boundary value problem the values of y and its derivatives at $x = x_0$ and $x = x_n$ are given, Eq. (7.2) is solved by iteration as explained below.

In order to start Eq. (7.2), several values of y, depending on the order of the differential equation (7.1), at $x = x_0$, $x = x_0 + \Delta x$, $x = x_0 + 2\Delta x$, $\ldots$, are needed. In boundary value problems not all of these values can be obtained from initial conditions given at $x = x_0$ since some of them should satisfy conditions at the other boundary $x = x_n$. In using iteration methods for the solution, all of the values of y for starting Eq. (7.2) are either taken from initial conditions at $x = x_0$ or are assigned trial values. Then Eq. (7.2) is applied successively up to solving for y_n, which is the value of y at the boundary x_n. If the values of $y_n, y_{n-1}, y_{n-2}, \cdots$, match the finite difference representation of the boundary condition at $x = x_n$ the solution is terminated; if not, the trial values of y for starting Eq. (7.2) are altered, usually linearly interpolated as a function of values of y at x_n boundary, and Eq. (7.2) is reapplied from $x = x_0$.

Method of Simultaneous Equations

This method involves the direct substitution of the differential equation by an equivalent finite-difference equation and the solution of the resulting simultaneous equations. The general equation representing y_{i+1} at $x_{i+1} = x_0 + i \Delta x$ (where x_0 is the boundary and i is the x index) is written in terms of previously available $y_i, y_{i-1}, y_{i-2}, \cdots$. The boundary conditions at two boundaries, $x = x_0$ and $x = x_n$, are also formulated in finite-difference form and incorporated with the general y_{i+1} equations. The final formulation should result in l equations in l unknowns, where l is a function of the number of increments of Δx in the region $x_0 \le x \le x_n$ and varies as a function of the order of the differential equation and the given boundary conditions. The methods of solution described above are used in the following example.

Example. It is required to find the solution of the differential equation

$$\frac{d^2 x}{dt^2} + 3x = t \tag{7.3}$$

with the initial and boundary conditions $(t = 0, x = 1)$ and $(t = 1, x = 1)$. The solution of this differential equation satisfying the given boundary conditions can be obtained from methods of the Appendix or texts on differential equations [1]. This solution is given as

$$x = \frac{(2/3) - \cos\sqrt{3}}{\sin\sqrt{3}} \sin\sqrt{3}\,t + \cos\sqrt{3}\,t + \frac{t}{3} \tag{7.4}$$

Equation (7.3) can be replaced by its difference form using central difference relations of Table 5.4 of Chapter 5. Since the solution is desired for the interval $t = 0$ to $t = 1$, Fig. 7-1(a), this interval is divided into n

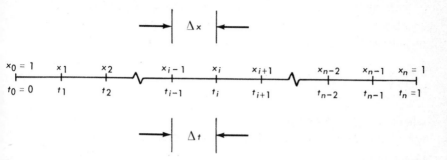

figure 7-1(a). *Increments for the solution of Eq. (7.3).*

equal parts resulting in $\Delta T = 1/n$. Note that the value of x at $t = 0$ is $x_0 = 1$ and at $t = 1$ is $x_n = 1$. Let us now replace the differential equation (7.3) with its central difference equivalent. For $t = t_1$

$$\frac{x_0 - 2x_1 + x_2}{\Delta t^2} + 3x_1 = t_1 \tag{7.5}$$

for $t = t_i (i = 2, 3, 4, \ldots, n - 2)$

$$\frac{x_{i+1} - 2x_i + x_{i+1}}{\Delta t^2} + 3x_i = t_i \tag{7.6}$$

for $t = t_{n-1}$

$$\frac{x_{n-2} - 2x_{n-1} + x_n}{\Delta t^2} + 3x_{n-1} = t_{n-1} \tag{7.7}$$

Iterative Method. The general form of Eq. (7.6) can be used to solve for x_{i+1} as

$$x_{i+1} = -x_{i-1} + x_i(2 - 3\Delta t^2) + (\Delta t)^2 t_i \tag{7.8}$$

$$\text{for } i = 1, 2, \ldots, n - 1$$

where $t_i = i \Delta t$.

Note that the values of x_0 and x_n are known from the initial ($t = 0$) and boundary ($t = 1$) conditions. The iterative method of solution starts by solving Eq. (7.8) for $i = 1$

$$x_2 = -x_0 + x_1 (2 - 3\Delta t^2) + (\Delta t)^2 t_1 \tag{7.9}$$

where $t_1 = \Delta t$. Since the value of $x_0 = 1$ is given in the problem, a value for x_1 should be assumed. Let $\Delta t = 0.2$; this will result in $n = 1/0.2 = 5$ increments. Using this value of Δt and assuming a trial value of $x_1 = 1$, Eq. (7.9) can be solved for x_2 and subsequently Eq. (7.8) can be solved for x_3, x_4, and x_5. This will result in

$$
\begin{aligned}
x_0 &= 1 \\
x_1 &= 1 \\
x_2 &= 0.888 \\
x_3 &= 0.68544 \\
x_4 &= 0.424627 \\
x_5 &= 0.144859
\end{aligned}
\tag{7.10}
$$

Since the value of $x_5 = x_n$ from the boundary condition should be 1.0 (higher than 0.144859 obtained above), let us assume another trial value for $x_1 = 2$. This results in

$$
\begin{aligned}
x_0 &= 1 \\
x_1 &= 2 \\
x_2 &= 2.768 \\
x_3 &= 3.21984 \\
x_4 &= 3.30930 \\
x_5 &= 3.03364
\end{aligned}
\tag{7.11}
$$

Now, we passed the boundary value at x_5. At this point we can use a linear interpolation between values of relations (7.10) and (7.11) to obtain the next trial value of x_1; that is,

$$x_1 = 1 + \left(\frac{2 - 1}{3.03364 - 0.144859}\right)(1 - 0.144859) \tag{7.12}$$

$$= 1.29602$$

where the parentheses on the right contain the desired value of x_n at the boundary and its value from relations (7.10). Using the above value of x_1 and reapplying Eqs. (7.9) and (7.8), we get

$$x_0 = 1$$

$$x_1 = 1.29602$$

$$x_2 = 1.44452$$

$$x_3 = 1.43568 \tag{7.13}$$

$$x_4 = 1.27855$$

$$x_5 = 1.0$$

which results in the desired value of boundary condition x_5.

Simultaneous Equations Method. Since the end conditions x_0 and x_n are given, x_1 through x_{n-1} are the unknowns. Setting the value of $x_0 = 1$ in Eqs. (7.5), writing Eq. (7.6) for $i = 2, 3, 4, \ldots, n - 2$, and setting the value of $x_n = 1$ in Eq. (7.7), we can write $n - 1$ equations in $n - 1$ unknowns x_1 through x_{n-1} as given in Table 7-1. For a given value of Δt the value of $n = 1/\Delta t$ can be calculated and the equations of Table 7-1 can be solved. These equations can be written in matrix notation as follows:

$$(A)(Q) = (Y) \tag{7.14}$$

where (A) is the square matrix of the coefficients of x_1 through x_{n-1}, (Q) is the column matrix of x's, and (Y) is the matrix of the constants given on the right side of Table 7-1. From Eq. (7.14), the solution (Q) can be obtained by premultiplying both sides of the expression by $(A)^{-1}$

$$(Q) = (A)^{-1}(Y) \tag{7.15}$$

Table 7-1. Finite Difference Representation of Eq. (7.3)

$$x_1(-2 + 3\Delta t^2) + x_2 \quad \text{-----------(For } t_1\text{)-----------} \quad = t_1(\Delta t)^2 - 1$$

$$x_1 \quad + x_2(-2 + 3\Delta t^2) + x_3 \text{---------(For } t_2\text{)---------} \quad = t_2(\Delta t)^2$$

$$+ x_2 \quad + x_3(-2 + 3\Delta t^2) + x_4 \quad = t_3(\Delta t)^2$$

$$\vdots$$

$$(\text{For } t_{n-2})\text{-------------} \quad x_{n-3} \quad + x_{n-2}(-2 + 3\Delta t^2) + x_{n-1} \quad = t_{n-2}(\Delta t)^2$$

$$(\text{For } t_{n-1})\text{----------} \quad x_{n-2} \quad + x_{n-1}(-2 + 3\Delta t^2) \quad = t_{n-1}(\Delta t) - 1$$

The solution was evaluated with a digital computer program to be discussed shortly. The results of the solution for various values of Δt are given in Table 7-2. It is noted that no significant change in the solution appears as the value of Δt is increased from 0.05 to 0.1 to 0.2. The analytical solution, evaluated from Eq. (7.4), is also given in Table 7-2.

Computer Program for the Solution of Eq. (7.3)

The following computer program has been written in BASIC programming language [2] for the numerical evaluation of the above solution. In the program, matrix methods are used to solve the simultaneous equations given in Table 7-1. In the BASIC programming language rows and columns of matrices are started from zero; thus the first element of the matrix is the (0, 0) element, and so on.

Table 7-2.   Results of the Solution of Eq. (7.3)

t	x From Eq. (7.4)	x $\Delta t = 0.05$	$\Delta t = 0.1$	$\Delta t = 0.2$
	1	1.0	1.0	1.0
.05	1.08541	1.08549		
.10	1.16281	1.16297	1.16338	
.15	1.23174	1.23198		
.20	1.29182	1.29212	1.29287	1.29602
.25	1.34271	1.34307		
.30	1.38416	1.38456	1.38556	
.35	1.41599	1.41643		
.40	1.32808	1.43854	1.43969	1.44453
.45	1.45039	1.45087		
.50	1.45295	1.45343	1.45462	
.55	1.44588	1.44635		
.60	1.42933	1.42979	1.43091	1.43568
.65	1.40358	1.40401		
.70	1.36893	1.36932	1.37028	
.75	1.32576	1.32611		
.80	1.27454	1.27483	1.27554	1.27856
.85	1.21576	1.21598		
.90	1.14999	1.15014	1.15053	
.95	1.07785	1.07792		
1.00	1.0	1.0	1.0	1.0

```
10    LET T=0.1
20    LET N=1/T
30    LET M=N-2
32    DIM A (25,25),Y(25,0),B(25,25),Q(25,0)
35    LET Z=-2 +3*T*T
40    MAT A=ZER(M,M)
45    MAT Y=ZER(M,0)
50    LET A(0,0)=Z
55    LET A(0,1)=1
60    LET Y(0,0)=T↑3-1
65    FOR J=1 TO M-1
70    LET A(J,J-1)=1
75    LET A(J,J)=Z
80    LET A(J,J+1)=1
85    LET Y(J,0)=(J+1)*T*(T↑2)
90    NEXT J
95    LET A(M,M-1)=1
100   LET A(M,M)=Z
105   LET Y(M,0)=(M+1)*T*(T↑2)-1
106   MAT PRINT A;
107   PRINT
108   MAT PRINT Y
110   MAT B=ZER(M,M)
115   MAT B=INV(A)
120   MAT Q=ZER(M,0)
125   MAT Q=B*Y
127   PRINT"ANSWER MATRIX Q"
128   PRINT
130   MAT PRINT Q
140   LET K=3↑.5
145   FOR D=0 TO 1 STEP T
150   LET X=(((2/3)-COS(K))/SIN(K))*SIN(K*D)+COS(K*D)+D/3
155   PRINT "D=  "D,"X=  "X
156   PRINT
160   NEXT D
170   END
```

Statement 10 of the program specifies the value of $\Delta t = T = 0.1$
Statement 20 computes the number of increments $N = 1/T = 10$. Since
Table 7-1 has $N - 1$ equations, the number of equations will be 9. As
mentioned previously, the matrix notation of BASIC language starts
from the zeroth row; this will mean that equations are numbered from
zero to 8. Or, by statement 30, $M = N - 2 = 8$. Statement 32 reserves
computer memory space for matrices used in the program. Statement

35 computes the diagonal coefficients of x_1, x_2, x_3, ... of Table 7-1. Statements 40 and 45 set the elements of matrices A and Y of Eq. (7.14) equal to zero. Statements 50 through 105 compute the elements of matrices A and Y as given in Table 7-1. Statements 106 through 108 print matrices A and Y. This is done for information only and, of course, does not affect succeeding results. The printout of these matrices is shown below.

$$
A = \begin{bmatrix}
-1.97 & 1 & 0 & 0 & 0 & 0 & 0 & 0 & 0 \\
1 & -1.97 & 1 & 0 & 0 & 0 & 0 & 0 & 0 \\
0 & 1 & -1.97 & 1 & 0 & 0 & 0 & 0 & 0 \\
0 & 0 & 1 & -1.97 & 1 & 0 & 0 & 0 & 0 \\
0 & 0 & 0 & 1 & -1.97 & 1 & 0 & 0 & 0 \\
0 & 0 & 0 & 0 & 1 & -1.97 & 1 & 0 & 0 \\
0 & 0 & 0 & 0 & 0 & 1 & -1.97 & 1 & 0 \\
0 & 0 & 0 & 0 & 0 & 0 & 1 & -1.97 & 1 \\
0 & 0 & 0 & 0 & 0 & 0 & 0 & 1 & -1.97
\end{bmatrix}
$$

$$
Y = \begin{bmatrix}
-0.999 \\
0.002 \\
0.003 \\
0.004 \\
0.005 \\
0.006 \\
0.007 \\
0.008 \\
-0.991
\end{bmatrix}
$$

Statement 110 sets elements of matrix B equal to zero. Statement 115 inverts matrix A and sets this inverse equal to B, that is, $B = A^{-1}$. Using this together with Eq. (7.15), the matrix containing the answers will be

$$Q = BY$$

given by statement 125. Statements 127 through 130 print out matrix Q. The rest of the statements — 140 through 160 — compute and print out the analytical solution given by Eq. (7.4). Note that statement 150 corresponds to algebraic Eq. (7.4). The results of this program are shown in Table 7-2.

From the solutions discussed above it can be seen that the iterative solutions are more adaptable to hand calculations and where computer routines for the solution of simultaneous equations are not available. From Table 7-1 it is seen that the boundary value problem can be reduced to the solution of a set of tridiagonal simultaneous equations. The recommended method of solution of this type of system as discussed in the next chapter is Thomas' method. However, if the number of equations is small, the use of other methods will not result in excessive computer time.

As was seen in the case of the foregoing example and as will be seen in the case of the following example, the iterative method of solution requires some judgment in "bracketing" the starting values (x_1 in the case of the previous problem). Translating this into a computer program may require an excessive number of logical operations.

Other Types of Boundary Conditions

In the above example the values of the function x were specified at the end points $t = 0$ and $t = 1$. In a number of problems the values of the derivatives of the function may be specified at the end points instead of the values of the function itself. Let us again consider Eq. (7.3) as repeated below:

$$\frac{d^2 x}{dt^2} + 3x = t \tag{7.16}$$

but this time with the boundary conditions ($t = 0$, $x = 1$) and ($t = 1$, $dx/dt = 1$). Note that in this case the derivative at $t = 1$ is specified. We again proceed with the formulation given in Fig. 7-1(a) and Eqs. (7.5),

7.6), and (7.7), but introduce the derivative at $t_n = 1$. Consider Fig. -1(b), which is the end boundary of the problem. In this case the ictitious point t_{n+1} is added. The value of the derivative at t_n can be /ritten using central differences

$$\left(\frac{dx}{dt}\right)_{t=t_n} = \frac{x_{n+1} - x_{n-1}}{2\Delta t} \tag{7.17}$$

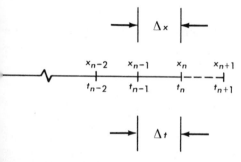

igure 7-1(b). *Boundary value at* $t_n = 1$.

Iterative Method. The solution of the problem in this case will pri-narily involve the application of Eq. (7.8) for $i = 1$ through n instead of $n - 1$ as was the case in the previous example. This allows the calcu-ation of x_{n+1}, which is required in the derivative Eq. (7.17). Note that rom the boundary condition the final value of the derivative should be unity

$$D = \frac{x_{n+1} - x_{n-1}}{2\Delta t} = 1 \tag{7.18}$$

Assuming a trial value for $x_1 = -1$ and using $\Delta t = 0.2$ as was done previously, Eq. (7.8) will result in

$$x_0 = 1$$
$$x_1 = -1$$
$$x_2 = -2.872$$
$$x_3 = -4.38336$$
$$x_4 = -5.34472$$
$$x_5 = -5.63271$$
$$x_6 = -5.20477$$
$$D = 0.349858$$

(7.19

Since this value is short of the boundary condition at x_5, we assum another trial value of $x_1 = -2$. This results in

$$x_0 = 1$$
$$x_1 = -2$$
$$x_2 = -4.752$$
$$x_3 = -6.91776$$
$$x_4 = -8.22939$$
$$x_5 = -8.52149$$
$$x_6 = -7.75101$$
$$D = 1.19594$$

(7.20

Using relations (7.19) and (7.20) and interpolating for the trial value x_1 for the next iteration, we get

$$x_1 = -1 + \left[\frac{-2 - (-1)}{1.19594 - 0.349858} \right](1 - 0.349858)$$

(7.21

$$= -1.76841$$

Using this value for the next trial, we get

$$x_0 = 1$$

$$x_1 = -1.76841$$

$$x_2 = -4.31662$$

$$x_3 = -6.33083$$

$$x_4 = -7.56134 \tag{7.22}$$

$$x_5 = -7.85249$$

$$x_6 = -7.16134$$

$$D = 0.999997$$

This solution satisfies the boundary and initial conditions.

Simultaneous Equations Method. Equating the slope from Eq. (7.17) to the given boundary condition $(dx/dt) = 1$ at $t = t_n = 1$, we get

$$x_{n+1} = 2\Delta t + x_{n-1} \tag{7.23}$$

The set of difference equations for this problem will be identical to the set given in Table 7-1 except for the last equation of the set and an additional equation for t_n. The last equation of Table 7-1 (for t_{n-1}) becomes

$$x_{n-2} + x_{n-1}(-2 + 3\Delta t^2) + x_n = t_{n-1}(\Delta t)^2 \tag{7.24}$$

Note that in the above equation x_n is an unknown. The equation for t_n can be written

$$x_{n-1} + x_n(-2 + 3\Delta t) + x_{n+1} = t_n(\Delta t)^2 \tag{7.25}$$

In this expression, the value of x_{n+1} can be substituted from Eq. (7.23), resulting in

$$2x_{n-1} + x_n(-2 + 3\Delta t) = t_n(\Delta t)^2 - 2\Delta t \tag{7.26}$$

Thus, the substitution of Eq. (7.24) in place of the equation for t_{n-1} of Table 7-1 and the addition of Eq. (7.26) for t_n will incorporate the new boundary condition in the problem. This will result in n equations in n unknowns, x_1 through x_n.

Use of Forward and Backward Differences in Incorporating Boundary Conditions

The boundary conditions can also be expressed in terms of forward and backward differences. Consider the second-order differential equation

$$y'' = \frac{d^2 y}{dx^2} = f(x, y, y')$$

(7.27)

with the boundary conditions $(x = x_0, dy/dx = y_0')$ and $(x = x_n, y' = y_n')$. Denoting the values of y at discrete x intervals as y_i and considering Fig. 7-2, Eq. (7.27) can be written in central difference form for point

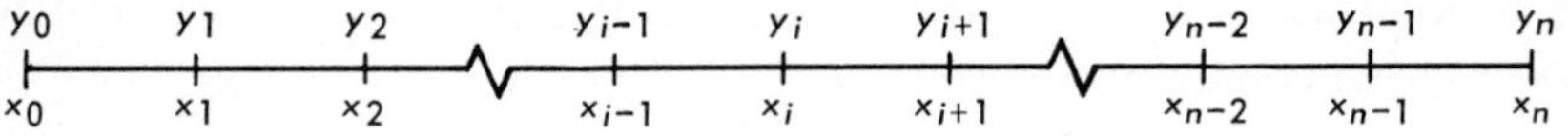

figure 7-2. *Use of forward and backward differences in boundary value problems.*

i. Using difference expressions from Table 5-4 of Chapter 5 for y'' and y',

$$\frac{y_{i+1} - 2y_i + y_{i-1}}{\Delta x^2} = f\left(x_i, y_i, \frac{y_{i+1} - y_{i-1}}{2\Delta x}\right)$$

(7.28)

Transposing the unknown y's to one side of the equation, we get

$$g_i(y_{i+1}, y_i, y_{i-1}, \Delta x) = w_i$$

(7.29)

where g_i is a function of the given variables and w_i is a function of known values of x_i. The above expression can be written for $i = 1$ to $i = n-1$ corresponding to values of y for x_2 through x_n

$$g_1(y_2, y_1, y_0, \Delta x) = w_1$$
$$g_2(y_3, y_2, y_1, \Delta x) = w_2$$
$$\dots\dots\dots\dots\dots\dots\dots\dots\dots$$

(7.30)

$$g_{n-2}(y_{n-1}, y_{n-2}, y_{n-3}, \Delta x) = w_{n-2}$$
$$g_{n-1}(y_n, y_{n-1}, y_{n-2}, \Delta x) = w_{n-1}$$

The above relations are $(n - 1)$ equations in $n + 1$ unknowns $y_0, y_1, \ldots, y_n$. The solution can be obtained only if two additional equations are given. These additional equations are derived from the boundary conditions.

The expression for the derivative of the function at y_0 in terms of the forward differences of Table 5-3 of Chapter 5 results in

$$\left(\frac{dy}{dx}\right)_{x = x_0} = Dy_0 = \frac{1}{2\Delta x}(-3y_0 + 4y_1 - y_2) \tag{7.31}$$

Equating this to the given value of the derivative at the boundary $x = 0$, we get

$$-3y_0 + 4y_1 - y_2 = 2(\Delta x)y_0' \tag{7.32}$$

The expression for the derivative of the function at y_n in terms of the backward differences of Table 5-2 of Chapter 5 results in

$$\left(\frac{dy}{dx}\right)_{x = x_n} = Dy_n = \frac{1}{2\Delta x}(3y_n - 4y_{n-1} + y_{n-2}) \tag{7.33}$$

Equating this to the given boundary condition at $x = x_n$, we get

$$3y_n - 4y_{n-1} + y_{n-2} = 2(\Delta x)y_n' \tag{7.34}$$

Equations (7.32) and (7.34) along with the set of equations (7.30) constitute $(n + 1)$ equations to be solved for $(n + 1)$ unknowns y_0 through y_n. Note that in the solution of Eq. (7.27) the differentials were always replaced by differences with errors of order h^2

Example. Consider the solution of the differential equation

$$y'' = y' + y + x^2 \tag{7.35}$$

where primes denote derivatives with respect to x, subject to the initial and boundary conditions $(x = 0, dy/dx = 1)$ and $(x = 1, dy/dx = 0)$. Replacing the differential equation with central differences of $e = 0(h^2)$ from Table 5-4 of Chapter 5, we get

$$\frac{y_{i+1} - 2y_i + y_{i-1}}{\Delta x^2} = \frac{y_{i+1} - y_{i-1}}{2\Delta x} + y_i + x_i^2 \tag{7.36}$$

Transposing the unknown y's to one side, we get

$$y_{i+1}(2 - \Delta x) + y_i(-4 - 2\Delta x^2) + y_{i-1}(2 + \Delta x) = x_i^2(2\Delta x^2) \tag{7.37}$$

Using Eq. (7.32) for the given derivative ($dy/dx = 1$) at $x = 0$, we get

$$-3y_0 + 4y_1 - y_2 = 2(\Delta x) \tag{7.38}$$

Using Eq. (7.34) for the given derivative ($dy/dx = 0$) at $x = 1$, we get

$$3y_n - 4y_{n-1} + y_{n-2} = 0 \tag{7.39}$$

Combining Eq. (7.38), Eq. (7.37) for $i = 1, 2, \ldots, (n-1)$ and Eq. (7.39), we get the set of equations

$$-3y_0 + 4y_1 - y_2 = 2\Delta x$$
$$(2 + \Delta x)y_0 + (-4 - 2\Delta x^2)y_1 + (2 - \Delta x)y_2 = 2(\Delta x)^2 x_1^2$$
$$(2 + \Delta x)y_1 + (-4 - 2\Delta x^2)y_2 + (2 - \Delta x)y_3 = 2(\Delta x)^2 x_2^2$$
$$\cdots\cdots\cdots\cdots\cdots\cdots\cdots\cdots\cdots\cdots\cdots\cdots\cdots\cdots\cdots\cdots \tag{7.40}$$
$$(2 + \Delta x)y_{n-2} + (-4 - 2\Delta x^2)y_{n-1} + (2 - \Delta x)y_n = x_{n-1}^2$$
$$y_{n-2} - 4y_{n-1} + 3y_n = 0$$

Equations (7.40) constitute a set of $(n + 1)$ equations in $(n + 1)$ unknown y_0 through y_n. In the above equations, $x_i = i\,\Delta x$. Equations (7.40) are solved for the unknowns for $\Delta x = 0.05$, $\Delta x = 0.1$, and $\Delta x = 0.2$. The results are given in Table 7-3.

Solution of Simultaneous Differential Equations

As mentioned in the previous chapter, a differential equation of nth order can be reduced to n first-order differential equations. Thus, a set

Table 7-3. Solution of Eq. (7.35)

x	$\Delta x = 0.5$	$\Delta x = 0.10$	$\Delta x = 0.20$
		y	
0	-2.11359	-2.11158	-2.10602
0.05	-2.06499		
0.10	-2.01918	-2.01718	
0.15	-1.97618		
0.20	-1.93598	-1.92297	-1.92823
0.25	-1.89858		
0.30	-1.86397	-1.86193	
0.35	-1.83214		
0.40	-1.80305	-1.80096	-1.79485
0.45	-1.77669		
0.50	-1.75301	-1.75085	
0.55	-1.73198		
0.60	-1.71352	-1.71126	-1.70448
0.65	-1.69760		
0.70	-1.68412	-1.68173	
0.75	-1.67302		
0.80	-1.66419	-1.66163	-1.65379
0.85	-1.65754		
0.90	-1.65294	-1.65018	
0.95	-1.65027		
1.00	-1.64938	-1.64636	-1.63690

of differential equations containing higher-order derivatives can be reduced to a set containing only first-order derivatives. The solution of
sets of first-order boundary value differential equations is obtained using
the procedures discussed above.

Consider the set of first-order differential equations

$$x' = f(x, y, t)$$
$$y' = g(x, y, t)$$

(7.41)

where primes denote derivatives with respect to t, with the initial and
boundary conditions $(t = 0, x = x_0)$ and $(t = 1, y = y_n)$. The above
expressions can be written in difference form using forward differences
of order h from Table 5-3 of Chapter 5. This results in

$$\frac{x_{i+1} - x_i}{\Delta t} = f(x_i, y_i, t_i)$$

$$\frac{y_{i+1} - y_i}{\Delta t} = g(x_i, y_i, t_i)$$

(7.42)

Collecting the unknowns x and y to one side, we get

$$h_i(x_{i+1}, x_i, y_i, \Delta t) = w_i$$

$$f_i(x_i, y_{i+1}, y_i, \Delta t) = z_i$$

(7.43)

where h and f are functions of the given variables and w and z are functions of the independent variables t and Δt. Writing these equations for $i = 0$ to $i = n - 1$ will result in $2n$ equations and $2(n + 1)$ unknowns; that is, x_0 through x_n and y_0 through y_n. Since the values of x_0 and y_n are given, these can be substituted into the equations, thereby reducing the number of unknowns to $2n$. At this point, the set of simultaneous equations can be solved by methods discussed in Chap. 1.

Example. Consider the second-order differential equation given in Eq. (7.35) and repeated below:

$$y'' = y' + y + x^2$$

(7.44)

with the boundary conditions $(x = 0, y' = 1)$ and $(x = 1, y' = 0)$. Denoting $y' = z$, Eq. (7.44) can be reduced to two first-order differential equations

$$y' = z$$

$$z' = z + y + x^2$$

(7.45)

In the case of these equations, the corresponding equations to Eq. (7.42) become

$$\frac{y_{i+1} - y_i}{\Delta x} = z_i$$

$$\frac{z_{i+1} - z_i}{\Delta x} = z_i + y_i + x_i^2$$

(7.46)

This results:

$$y_{i+1} - y_i - \Delta x z_i = 0$$
$$z_{i+1} + z_i(-1 - \Delta x) - \Delta x y_i = \Delta x (x_i)^2$$

(7.47)

Dividing the interval $0 \leq x \leq 1$ into n equal increments Δx, Eqs. (7.47) can be written for $i = 0$ to $i = n - 1$ for the unknowns y_0 to y_n and z_0 to z_n

$$
\begin{array}{llll}
-y_0 \quad +y_1 & & & = \Delta x z_0 \\
\quad -y_1 \quad +y_2 & \quad -\Delta x z_1 & & = 0 \\
\quad\quad -y_2 \quad +y_3 & \quad\quad -\Delta x z_2 & & = 0 \\
\quad\quad\quad -y_{n-1} \quad +y_n & \quad\quad\quad -\Delta x z_{n+1} & & = 0 \\
-\Delta x y_0 \quad\quad +z_1 & & & = (1 + \Delta x) z_0 + (\Delta x) x_0 \\
\quad -\Delta x y_1 \quad -(1 + \Delta x) z_1 +z_2 & & & = \Delta x (x_1)^2 \\
\quad\quad -\Delta x y_2 & & & \\
\quad\quad\quad \cdots & & & \\
\quad\quad\quad -\Delta x y_{n-1} \quad -(1 + \Delta x) z_{n-1} & = \Delta x (x_{n-1})^2 - z_n &
\end{array}
$$

(7.48)

Setting the values $z_0 = 1$ and $z_n = 0$ in the above equations from the given boundary conditions, we will obtain $2n$ equations in $2n$ unknowns y_0 through y_n and z_1 through z_{n-1}.

The solutions of these equations are obtained for $\Delta x = 0.1$ and $\Delta x = 0.2$ corresponding to $n = 1/0.1 = 10$ and $n = 1/0.2 = 5$ intervals between $x = 0$ to $x = 1.0$. These results are presented in Table 7-4, and can be compared to the solution given in Table 7-3. Note that the solution of Eqs. (7.48) also gives the values of $z = dy/dx$ in addition to the values of y. The difference between values of y between Tables 7-3 and 7-4 comes primarily from the fact that the formulation of Eq. (7.46) contains an error of order h while formulation of Eq. (7.36) contains an error of order h^2.

Table 7-4. Solution of Eqs. (7.48)

x	$\Delta x = 0.1$		$\Delta x = 0.2$	
	Y	Z	Y	Z
0	-2.11814	1.0*	-2.11859	1.0*
0.1	-2.01814	0.888186		
0.2	-1.92932	0.776191	-1.91859	0.776283
0.3	-1.85170	0.664878		
0.4	-1.78521	0.555195	-1.76333	0.555822
0.5	-1.72969	0.448193		
0.6	-1.68488	0.345043	-1.65217	0.346320
0.7	-1.65037	0.247060		
0.8	-1.62566	0.155729	-1.58290	0.157150
0.9	-1.61009	0.072735		
1.0	-1.60282	0.0*	-1.55147	0.0*

*Inserted from boundary conditions.

STABILITY OF SOLUTION OF DIFFERENTIAL EQUATION

As discussed here and in the previous chapter, the numerical solutions of differential equations depend on the particular finite-difference representation of the equation and the increment of the independent variable, Δx or Δt. Correlating the stability and the accuracy of the solution of the difference representation with the size of the increment is an important step toward the justification of the finite-difference representation (see stability discussion of the previous chapter). Unfortunately, this correlation can be obtained analytically in a limited number of cases. In many practical applications the correlation of Δx and Δt with the accuracy of solution is made by running particular finite difference formulations on the computer and examining the resulting solutions.

Analytic methods of stability and accuracy determination of finite-difference formulations involve the use of *difference-differential* equations, as described below.

ANALYTIC SOLUTION OF DIFFERENCE-DIFFERENTIAL EQUATIONS

The finite-difference formulations of differential equations can be represented in the form

$$x(t + iT) + v_1 x[t + (i - 1) T] + v_2 x[t + (i - 2) T]$$
$$+ \cdots + v_i x(t) = f(t) \tag{7.49}$$

where T is the increment in t and i is an integer. Using the previously given notation

$$E^i x(t) = x(t + iT) \tag{7.50}$$

Eq. (7.49) becomes

$$\left(E^i + v_1 E^{i-1} + v_2 E^{i-2} + \cdots + v_i\right) x(t) = f(t) \tag{7.51}$$

Following [3], the *complementary* solution of the above equation x_c can be obtained by solving

$$\left(E^i + v_1 E^{i-1} + v_2 E^{i-2} + \cdots + v_i\right) x_c(t) = 0 \tag{7.52}$$

The solution of Eq. (7.52) is obtained by assuming

$$x_c(t) = Ce^{mt} \tag{7.53}$$

where C is an arbitrary constant, and m is a number to be determined. Substituting Eq. (7.53) in Eq. (7.52), we have

$$Ce^{mt}\left(e^{miT} + v_1 e^{m(i-1)T} + v_2 e^{m(i-2)T} + \cdots + v_i\right) = 0 \tag{7.54}$$

Dividing through by Ce^{mt} and substituting $q = e^{mT}$, we get

$$q^i + v_1 q^{i-1} + v_2 q^{i-2} + \cdots + v_i = 0 \tag{7.55}$$

(Note that this equation is identical to Eq. (7.52) with q's replacing E's.) Assuming that $t = nT$, where n is an integer, the solution (7.53) becomes

$$x_c(nT) = Ce^{mnT} = Cq^n \tag{7.56}$$

The roots of Eq. (7.55) will give the values of q to be used in the solution equation (7.56). The roots of Eq. (7.55) can be obtained by standard polynomial root finding methods discussed in Chap. 3. Three cases are encountered after the roots are obtained. They are:

  1. *Distinct Roots.* In which case the solution (7.56) becomes

$$x_c(nT) = C_1 q_1{}^n + C_2 q_2{}^n + C_3 q_3{}^n + \cdots + C_i q_i{}^n \qquad (7.57)$$

where $q_1, q_2, \ldots, q_i$ represent the roots of the polynomial (7.55).

  2. *Complex Roots.* In this case the roots will be complex conjugate as the coefficients of Eq. (7.55) are real (see Chap. 3). The conjugate roots q_1, q_2 will be in the form

$$q_1 = a + bj \qquad q_2 = a - bj \qquad (7.58)$$

where a and b are constants and $j^2 = -1$. These roots can be written as

$$q_1 = Re^{j\theta} \qquad q_2 = Re^{-j\theta} \qquad (7.59)$$

where $R = \sqrt{a^2 + b^2}$ and $\theta = \tan^{-1}(b/a)$. The corresponding solution may be expressed

$$\begin{aligned} x_c(nT) &= C_1 q_1{}^n + C_2 q_2{}^n \\ &= R^n\left(C_1 e^{jn\theta} + C_2 e^{-jn\theta}\right) \end{aligned} \qquad (7.60)$$

Substituting $e^{\pm jn\theta} = \cos n\theta \pm j \sin n\theta$ in the above equation, the solution becomes

$$x_c(nT) = R^n(C_1' \cos n\theta + C_2' \sin n\theta) \qquad (7.61)$$

where C_1' and C_2' are new constants.

  3. *Repeated Roots.* For repeated roots $q_1 = q_2 = q$ the solution is expressed as

$$x_c(nT) = q^n(C_1 + C_2 n) \qquad (7.62)$$

The *particular* solution of Eq. (7.49), x_p, can be obtained in analogy with the differential equation theory [1]. In obtaining particular solutions we assume several types of solution, depending on the form of the function $f(t)$, in Eq. (7.51). This solution is then substituted in Eq. (7.51) and the constants are evaluated. The following types of $f(t)$ are often encountered in practice.

1. $f(t) = A^{kt} = A^{knT}$ where A and k are constants. In this case a particular solution of the type

$$x_p(nT) = CA^{knT} \tag{7.63}$$

is assumed and this solution is substituted in Eq. (7.51). The resulting equation is then solved for the constant C. Thus, the complete solution of the differential equation becomes

$$x = x_c(nT) + x_p(nT) \tag{7.64}$$

2. $f(t) = At^m = A(nT)^m$ where A is a constant and m is an integer. In this case, a solution of the type

$$x_p(nT) = C_1(nT)^m + C_2(nT)^{m-1} + \cdots + C_{m+1} \tag{7.65}$$

is assumed and this solution is substituted in the difference equation (7.51). The values of $C_1, C_2, \ldots, C_{m+1}$ are computed by equating the coefficients of equal powers of n on both sides of the resulting equation.

3. $f(t) = \sin AT$ or $\cos AT = \sin AnT$ or $\cos AnT$. In this case the exponential equivalent of these functions can be written and the solution can be obtained by the method of case 1 above. Alternatively, a solution of the type

$$x_p(nT) = C_1 \sin AnT + C_2 \cos AnT \tag{7.66}$$

can be assumed and substituted in Eq. (7.51). The values of the constants C_1 and C_2 are obtained by equating the coefficients of $\sin AnT$ and $\cos AnT$ terms on both sides of the equation, respectively.

From the solutions discussed above it can be seen that the complementary solution (7.56) represents the transient part of the solution as q's are raised to the number of the time step n. Thus, since q's in finite difference formulations are usually functions of the time increment T, the complementary solution may be *unstable* for some region of T values. On the other hand, the *particular* solution follows the "driving input" of the difference equation as seen from the three cases above. For this reason the stability of a specific finite difference solution may be determined from an examination of the complementary solution of its difference-differential equation. The regions of T values for a stable solution are chosen by an examination of the complementary solution of the difference equation.

Example. Consider the error analysis of the finite-difference solution of the differential equation

$$\frac{dy}{dt} + 2y = 3t \tag{7.67}$$

with the initial condition $t = 0$, $y = 1$.

The analytic solution of the equation can be obtained by standard methods of solution of differential equations. See Appendix on differential equations. The *complementary* solution is the solution satisfying

$$\frac{dy}{dt} + 2y = 0$$

or

$$(D + 2)y = 0 \tag{7.68}$$

where D is the derivative operator. The root of this equation is $D = -2$ and the complementary solution becomes

$$y = Ce^{-2t} \tag{7.69}$$

The *particular* solution is obtained by assuming a solution of the type

$$y_p = At + B \tag{7.70}$$

and substituting in Eq. (7.67)

$$A + 2At + 2B = 3t \tag{7.71}$$

Equating like powers of t, we have

$$A = \frac{3}{2} \qquad B = -\frac{3}{4}$$

The total analytic solution of the differential equation (7.67) becomes

$$y = Ce^{-2t} + \frac{3}{2}t - \frac{3}{4} \tag{7.72}$$

Setting the initial condition $t = 0$, $y = 1$ in Eq. (7.72), we obtain $C = 7/4$. Or Eq. (7.72) with this value of C becomes

$$y = \frac{7}{4}e^{-2t} + \frac{3}{2}t - \frac{3}{4} \tag{7.73}$$

which is the complete analytic solution of differential equation (7.67). Equation (7.67) can be represented in finite-difference form as

$$\frac{y_{n+1} - y_n}{T} + 2y_n = 3t_n = 3nT \tag{7.74}$$

where T is the increment in t, that is, $T = \Delta t$. Denoting $2T - 1 = a$, Eq. (7.74) becomes

$$(E + a)y_n = 3(nT)T \tag{7.75}$$

The complementary solution is obtained by setting the right side of this equation equal to zero:

$$(E + a)y_n = 0 \tag{7.76}$$

The root of this equation is $q = -a$, and the complementary solution becomes

$$(y_n)_{comp} = C(-a)^n \tag{7.77}$$

From the above complementary solution it can be seen that for large n and $a \geq 1$ the solution will grow beyond bounds. Therefore, the complementary solution will be stable only for

$$a = 2T - 1 < 1$$

or

$$T < 1 \tag{7.78}$$

The particular solution of Eq. (7.75) is obtained by assuming a particular solution of the type

$$\left(y_n\right)_{\text{part}} = A(nT) + B \tag{7.79}$$

and substituting in Eq. (7.75). This will result in

$$A(n + 1)T + B + aA(nT) + aB = 3(nT)T \tag{7.80}$$

Equating the coefficients of like powers of n on both sides of the equation, we get

$$A = \frac{3T}{1 + a} = \frac{3}{2}$$
$$B = \frac{3T^2}{(1 + a)^2} = -\frac{3}{4} \tag{7.81}$$

In the above, the substitution $1 + a = 2T$ is used. Thus, the complete finite-difference solution of difference equation (7.74) will be the sum of Eqs. (7.77) and (7.79) with the constants of Eq. (7.81)

$$y_n = y(nT) = C(-a)^n + \frac{3}{2}(nT) - \frac{3}{4} \tag{7.82}$$

The constant C is evaluated by using the initial condition $t = nT = 0$, $y = 0$, or $n = 0$, $y = 0$. Using this in Eq. (7.82) will result in $C = 7/4$, or

$$y(nT) = \frac{7}{4}(-a)^n + \frac{3}{2}(nT) - \frac{3}{4} \tag{7.83}$$

Note that $-a = 1 - 2T$. Using this relation in the above, and subtracting Eq. (7.83) from analytic solution (7.73), we get (set $t = nT$ in Eq. (7.73)):

$$\epsilon = y(t) - y(nT) = \frac{7}{4}[e^{-2nT} - (1 - 2T)^n] \qquad (7.84)$$

where ϵ is the error between the analytic and finite-difference solutions. Note that the *particular* solution does not affect the error term as it is canceled by the corresponding analytic solution. The exponential function in (7.84) can be expanded in Taylor series as follows:

$$\left(e^{-2T}\right)^n = \left(1 - 2T + \frac{(2T)^2}{2!} - \frac{(2T)^3}{3!} + \cdots\right)^n \qquad (7.85)$$

Thus, the effective error of Eq. (7.84) is introduced by neglecting the second and higher powers of T in the difference solution represented by $(1 - 2T)^n$. The values of percentage error (ϵ/y) (100) (using Eqs. (7.73) and (7.74)) were evaluated for various values of t using T as a parameter. A plot of this is given in Fig. 7-3.

Example. As an example of the above analysis for a second-order differential equation, consider the differential equation

$$\frac{d^2 x}{dt^2} + \omega^2 x = 0 \qquad (7.86)$$

with the initial conditions $t = 0$, $x = 1$, and $dx/dt = 0$. The analytic solution of Eq. (7.86) is of the form

$$x = A \sin \omega t + B \cos \omega T \qquad (7.87)$$

where A and B are constants. The initial conditions will establish the values of A and B as follows

$$A = 0 \qquad B = 1$$

Using these, solution (7.87) becomes

$$x(t) = x(nT) = \cos \omega t = \cos n(\omega T) \qquad (7.88)$$

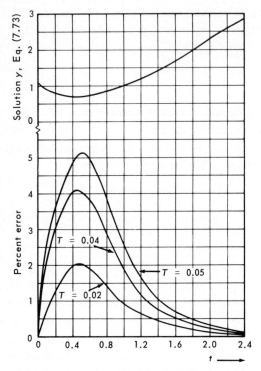

figure 7-3. *Error in difference formulation and the solution of Eq. (7.67) vs. t.*

where $t = nT$ (n is an integer, and T is the time step) is substituted for calculations to be performed later.

The difference representation of Eq. (7.86) using central differences of order T^2 can be written

$$\frac{x_{n+1} - 2x_n + x_{n-1}}{T^2} + \omega^2 x_n = 0 \qquad (7.89)$$

Shifting the index n by one unit, the above equation can be written

$$x_{n+2} + x_{n+1}(\omega^2 T^2 - 2) + x_n = 0 \qquad (7.90)$$

In analogy to Eq. (7.52), the above becomes

$$[E^2 + (\omega^2 T^2 - 2)E + 1]x_n = 0 \qquad (7.91)$$

The roots of this equation are

$$q_1, q_2 = \frac{-(\omega^2 T^2 - 2) \pm \sqrt{(\omega^2 T^2 - 2)^2 - 4}}{2}$$ (7.92)

Considering the part under the radical, three cases will arise

Case 1. $(\omega^2 T^2 - 2)^2 > 4$ or $\omega T > 2$ the roots q_1 and q_2 will be real valued and distinct. The solution will be of the type given by Eq. (7.57):

$$x(nT) = C_1 q_1{}^n + C_2 q_2{}^n$$ (7.93)

This is clearly a nonoscillatory solution and is unacceptable since the analytic solution (7.88) is oscillatory.

Case 2. $(\omega^2 T^2 - 2) = 4$ or $\omega T = 2$ double root $q_1 = q_2 = q$. The solution will be of the type given by Eq. (7.62):

$$x(nT) = q^n(C_1 + C_2 n)$$ (7.94)

This is again clearly a nonoscillatory solution, and is unacceptable for the same reason as in case 1.

Case 3. $(\omega^2 T^2 - 2)^2 < 4$ or $\omega T < 2$ complex conjugate roots of the type given by Eq. (7.58)

$$\begin{aligned} q_1 &= a + b_j \\ q_2 &= a - b_j \end{aligned}$$ (7.95)

with $a = -(\omega^2 T^2 - 2)/2$ and $b = (1/2)\sqrt{4 - (\omega^2 T^2 - 2)^2}$. The solution will be of the type given by Eq. (7.61)

$$x(nT) = R^n(C_1' \cos n\theta + C_2' \sin n\theta)$$ (7.96)

with

$$R = a^2 + b^2 = 1.0$$

$$\theta = \tan^{-1} \frac{b}{a}$$ (7.97)

The constants C_1' and C_2' can be obtained using the initial conditions for $t = nT = 0$ or $n = 0$. For $n = 0$ and $x = 1$, Eq. (7.96) will give $C_1' = 1$. The initial condition on the derivative dx/dt at $t = 0$ can be incorporated by writing the difference equation for the derivative as

$$\frac{x[(n + 1)\, T] - x\,[(n - 1)\, T]}{2T} = 0 \qquad (7.98)$$

or

$$x[(n + 1)\, T] = x[(n - 1)\, T] \text{ at } n = 0$$

or using Eq. (7.96) with $R = 1$,

$$C_1' \cos(n + 1)\theta + C_2' \sin(n + 1)\theta$$
$$= C_1' \cos(n - 1)\theta + C_2' \sin(n - 1)\theta \qquad (7.99)$$

Setting $n = 0$ and equating the coefficients of sine and cosine terms, respectively, we get

$$C_2' = 0 \qquad (7.100)$$

Thus the solution of difference equation (7.90) with the appropriate initial conditions becomes

$$x(nT) = \cos n\theta \qquad (7.101)$$

where θ is given by Eq. (7.97).

Comparisons of solutions (7.101) and (7.88) will show that the value of T should be so chosen as to make $\theta \simeq \omega T$. Note that ω is given in the problem equation (7.86). Using Eq. (7.97) together with the values of a and b in terms of ωT, several values of θ as a function of ωT were calculated as given in the following table. From this table it is noted that values of T corresponding to ωT of up to 0.2 seem acceptable.

ωT	θ
0	0
.1	.100042
.2	.200335
.3	.301137
.4	.402716
.5	.50536
.6	.609385
.7	.715142
.8	.823034
.9	.933531
1.	1.0472
1.1	1.16473
1.2	1.287

PROBLEMS

1. Formulate the solution of the differential equation

$$\frac{d^2 x}{dt^2} + 5 \frac{dx}{dt} + 3x = t^2$$

   subject to the initial and boundary conditions $(t = 0, x = 1)$ and $(t = 2, dx/dt = 3)$, by using central differences of order T^2 and dividing the interval of $0 \le t \le 2$ into n equal parts.

2. Formulate Prob. 1 with the boundary conditions $(t = 0, dx/dt = 0)$ and $(t = 2, dx/dt = 1)$.

3. Formulate Prob. 2 using forward and backward differences to represent the boundary conditions.

4. Formulate the solution of the differential equation

$$y'' = \frac{d^2 y}{dx^2} = 12x^2 + \frac{3}{x^2}$$

   subject to the condition $y' = y = 0$ when $x = 1$ for $0 \le x \le 1$.

5. Formulate the solution of the differential equation

$$\frac{d^2 y}{dx^2} - \frac{dy}{dx} = x$$

subject to the condition $y = 1$, $dy/dx = 0$ when $x = 1$ for $0 \le x \le 1$.

6. Obtain the solution of the boundary value differential equation

$$\frac{d^2 y}{dx^2} + 10 \frac{dy}{dx} + 16x = 0$$

with $y = 1$ at $x = 0$ and $y = 2$ at $x = 1$, by the iteration method.

7. Obtain the solution of the differential equation

$$\frac{d^2 y}{dx^2} = 12x^2 + \frac{3}{x^2}$$

with $dy/dx = y = 0$ at $x = 1$, by the iteration method.

8. Solve Prob. 7 for boundary condition $y = 1$ at $x = 0$ and $y = 2$ at $x = 1$.

9. Obtain the solution of Eq. (7.16) by the simultaneous equations method and compare with results given by relations (7.22).

10. Obtain the solutions given for $\Delta x = 0.2$ of Table 7-3.

11. Formulate the solution of the sets of differential equations

$$\frac{dx}{dt} = -\frac{y}{t} + \frac{x}{t} + 1$$

$$\frac{dy}{dt} = -\frac{x}{t} + \frac{y}{t} + 1$$

subject to the conditions $x = 0$ when $t = 0$, $y = 1$ when $t = 1$.

12. Solve the difference-differential equation

$$(E^2 - 5E + 6) y = e^t$$

13. Solve the difference-differential equation

$$(E^2 - 1)y = t^2 + 2t$$

subject to the conditions $y = 0$ when $t = 0$ and $y = 1$ when $t = 1$.

4. Solve the difference-differential equation

$$(E^2 - 2E + 4)y = x^2$$

subject to the conditions $y = 0$ when $x = 0$ and $dy/dx = 1$ when $x = 0$.

5. Discuss the stability of the solution of the central difference formulation of

$$\frac{d^2 y}{dx^2} + 2\frac{dz}{dx} + 3y = x^2$$

by examining the complementary solution of the difference-differential equation.

6. Discuss the stability of solution of the central difference representation of the equation

$$\frac{dy}{dx} + y = x^2$$

using the complementary solution of difference-differential equation.

7. Discuss the accumulation of error in the central difference formulation of the differential equation

$$\frac{d^2 y}{dx^2} = x$$

subject to the conditions $y = 0$, $dy/dx = 1$ when $x = 0$. The analytical solution of the above equation is

$$y = \frac{1}{6}x^3 + C_1 x + C_2$$

18. Discuss the accumulation of error in the forward difference formulation of the differential equation

$$\frac{dx}{dt} = x$$

subject to the condition $x = 5$ when $t = 0$. The analytical solution of the problem is

$$x = Ce^t$$

19. Discuss the accumulation of error in the forward difference formulation of the differential equation

$$\frac{dy}{dx} + y = e^x$$

subject to the conditions $y = 1$ when $x = 0$. The analytical solution of the problem is

$$y = (x + C)e^x$$

REFERENCES

1. Boyce, William E., and Richard C. DiPrima: "Elementary Differential Equations and Boundary Value Problems," John Wiley & Sons, Inc., New York, 1969.
2. BASIC Language, reference manual, General Electric Company, May, 1966.
3. Jordan, C.: "Calculus of Finite Differences," chap. 11, Chelsea Publishing Co., New York, 1947.
4. Todd, John (ed.): "Survey of Numerical Analysis," McGraw-Hill Book Company, New York, 1962.
5. Henrici, Peter: "Elements of Numerical Analysis," John Wiley & Sons, Inc., New York, 1964.

8 NUMERICAL SOLUTION OF PARTIAL DIFFERENTIAL EQUATIONS

The numerical solution of partial differential equations can be a subject of extensive study by itself. There are several excellent texts and references written on this subject, including [1, 2]. The presentation of the subject given in the following pages should be considered only as an introduction rather than as a complete treatment.

The discussions start by giving simple examples to illustrate the general concepts. The generalized forms of the solutions and addition methods of solutions are given after the introductory material.

A second-order partial differential equation can be written in the form

$$A(x, y) \frac{\partial^2 u}{\partial x^2} + B(x, y) \frac{\partial^2 u}{\partial x \partial y} + C(x, y) \frac{\partial^2 u}{\partial y^2} + f\left(x, y, u, \frac{\partial u}{\partial x}, \frac{\partial u}{\partial y}\right) = 0$$

$$(8.1)$$

where A, B, C, and f are functions of the given variables. If f is a *linear* function of the variables in parantheses, the partial differential equation (8.1) is considered to be a *linear partial differential equation*. If f is a *nonlinear* function of the given variables, the partial differential equation (8.1) is considered to be a *quasi-linear* equation since the second-order derivatives $\partial^2 u/\partial x^2$, $\partial^2 u/\partial x \partial y$, ..., appear in the first power only. The partial differential equation (8.1) is categorized depending on the relationship of A, B, and C coefficients as follows:

$B^2 - 4AC < 0$ elliptic equation

$B^2 - 4AC = 0$ parabolic equation

$B^2 - 4AC > 0$ hyperbolic equation

Examples of the above categories, often encountered in practice, are as follows:

Laplace's equation $\dfrac{\partial^2 u}{\partial x^2} + \dfrac{\partial^2 u}{\partial y^2} = 0$ elliptic

Heat equation $\dfrac{\partial^2 u}{\partial x^2} = a \dfrac{\partial u}{\partial t}$ parabolic

(where a is a constant and x and t are space and time coordinates).

Wave equation $c^2 \dfrac{\partial^2 u}{\partial x^2} = \dfrac{\partial^2 u}{\partial t^2}$ hyperbolic

(where c is a constant and x and t are space and time coordinates).

The above partial differential equations are written in the cartesian (x, y) coordinate system. These equations can also be written in other coordinate systems such as cylindrical and spherical. The following discussion, however, will be limited to cartesian coordinate system representations.

PARTIAL DIFFERENCE OPERATORS IN CARTESIAN COORDINATE SYSTEM

Consider the function

$$z = f(x, y) \tag{8.2}$$

where x, y are the cartesian coordinate system (Fig. 8-1), and the partial derivatives are denoted by

$$D_x = \frac{\partial z}{\partial x} \quad D_y = \frac{\partial z}{\partial y} \quad D_{x,y} = \frac{\partial^2 z}{\partial x \partial y} \quad D^2 x = \frac{\partial^2 z}{\partial x^2} \quad D^2 y = \frac{\partial^2 z}{\partial y^2} \tag{8.3}$$

Taking the point (0, 0) as the origin the partial derivatives of Eqs. (8.3) become (using central differences)

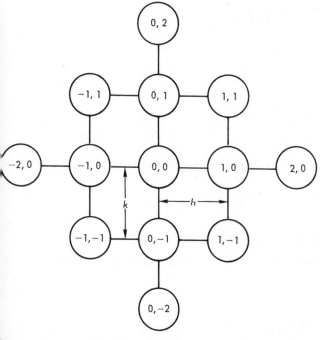

figure 8-1. *Cartesian coordinate system.*

$$2h \, D_x z_0 = z_{1,0} - z_{-1,0}$$

$$h^2 \, D_x^2 z_0 = z_{1,0} - 2z_{0,0} + z_{-1,0} \tag{8.4}$$

$$2h^3 \, D_x^3 z_0 = z_{2,0} - 2z_{1,0} + 2z_{-1,0} - z_{-2,0}$$

where h is the increment in the x direction.

Similarly, denoting k the increment in the y direction, we have

$$2k \, D_y z_0 = z_{0,1} - z_{0,-1}$$

$$k^2 \, D_y^2 z_0 = z_{0,1} - 2z_{0,0} + z_{0,-1} \tag{8.5}$$

$$2k^3 \, D_y^3 z_0 = z_{0,2} - 2z_{0,1} + 2z_{0,-1} - z_{0,-2}$$

The expression for the mixed derivative D_{xy} of the function z is obtained by applying the derivative operator D_y to the existing expression for D_x. Thus

$$D_{xy} z_0 = D_y (D_x) = D_y \left(\frac{z_{1,0} - z_{-1,0}}{2h} \right)$$

$$= \frac{(z_{1,1} - z_{1,-1}) - (z_{-1,1} - z_{-1,-1})}{4kh} \tag{8.6}$$

$$= \frac{z_{1,1} - z_{1,-1} - z_{-1,1} + z_{-1,-1}}{4kh}$$

The above expression in analogy to Eqs. (8.4) and (8.5) can be written

$$4kh \, D_{xy} z_0 = z_{1,1} - z_{1,-1} - z_{-1,1} + z_{-1,-1} \tag{8.7}$$

SOLUTION OF LAPLACE'S EQUATION

The Laplace partial differential equation, as previously given, can be written in cartesian coordinates as

$$\nabla^2 u = \frac{\partial^2 u}{\partial x^2} + \frac{\partial^2 u}{\partial y^2} = 0 \tag{8.8}$$

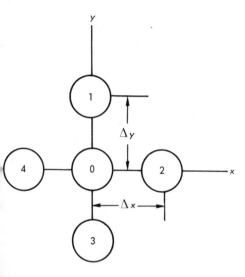

figure 8-2. *Solution of Laplace's equation.*

where ∇^2 is the *Laplace operator* (not to be confused with backward difference operator). Considering Fig. 8-2 and using difference relations given by Eqs. (8.4) and (8.5) in Eq. (8.8), we get

$$\frac{u_2 - 2u_0 + u_4}{\Delta x^2} + \frac{u_1 - 2u_0 + u_3}{\Delta y^2} = 0 \qquad (8.9)$$

Setting the values of $\Delta x = \Delta y = h$, Eq. (8.9) becomes

$$u_1 + u_2 + u_3 + u_4 - 4u_0 = 0 \qquad (8.10)$$

Using the above relation and assuming the values of u's as unknowns, the solution of Eq. (8.8) can be obtained by solving a series of simultaneous equations as illustrated in the following example:

Example. Obtain the solution of Laplace's equation for the geometry of Fig. 8-3. In this figure the values of u at the upper and lower boundary are set at 1,000, while the values of u at the right and left boundaries are set at zero. The values at the corners are arbitrarily set at 500 so as to make the discontinuity of approaching the corners from the x and y directions equal, i.e., the change from the nearest point to the corner

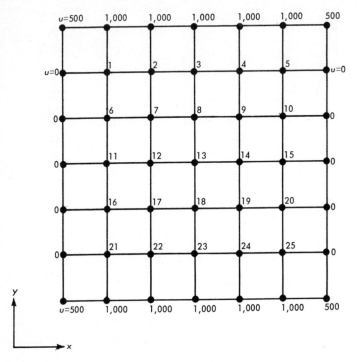

figure 8-3. *Example problem. Solution of Laplace's equation.*

from the x and y directions is 500. The square cross section is divided into six equal increments of length h in both x and y directions. This will give 25 points or 25 values of u to be calculated. Since the divisions of the cross section are symmetrical, only the values of u at the following points will be calculated:

$$u_1, \, u_2, \, u_3, \, u_6, \, u_7, \, u_8, \, u_{11}, \, u_{12}, \text{ and } u_{13}$$

Using Eq. (8.10), equations can be written for these unknowns. These equations are given in Table 8-1. Note that in obtaining the coefficients listed in the table the following relations, arising from the symmetry of the problem, were used:

$$u_2 = u_4 \quad u_7 = u_9 \quad u_6 = u_{16} \quad u_7 = u_{17} \quad u_8 = u_{18} \quad u_{12} = u_{14}$$

The solutions of these equations by previously outlined methods of Chap. 1 were evaluated using a digital computer. The answers are

Table 8-1. Coefficients of Simultaneous Equations

Point \ Unknown →	u_1	u_2	u_3	u_6	u_7	u_8	u_{11}	u_{12}	u_{13}	Constant
1	-4	1		1						-1,000
2	1	-4	1		1					-1,000
3		2	-4			1				-1,000
6	1			-4	1		1			0
7		1		1	-4	1		1		0
8			1		2	-4			1	0
11				2			-4	1		0
12					2		1	-4	1	0
13						2		2	-4	0

as follows:

$$u_1 = 500.0$$

$$u_2 = 682.692$$

$$u_3 = 730.769$$

$$u_6 = 317.308$$

$$u_7 = 500.0 \qquad\qquad (8.11)$$

$$u_8 = 557.692$$

$$u_{11} = 269.231$$

$$u_{12} = 442.308$$

$$u_{13} = 500.0$$

Note that these values represent the steady-state temperature distribution in a homogeneous solid material if u is considered to be the temperature.

ITERATIVE METHOD OF SOLUTION OF LAPLACE'S EQUATION

The solution of Laplace's equation (8.8) can be obtained by a simple iterative method due to Liebmann. This method is based on the fact that the value of u at any point is the average of the four surrounding values as given by Eq. (8.10), that is,

$$u_0 = \frac{1}{4}(u_1 + u_2 + u_3 + u_4) \tag{8.12}$$

The method starts by assuming an initial distribution of u over the region and calculating new values at each iteration using Eq. (8.12). The procedure is further explained through the following example.

Example. Again consider Fig. 8-3 of the previous example, but this time let us use matrix notation to specify various points of the cross section, as given in Fig. 8-4. Using the same conditions at the boundaries as given in the previous example, and assuming a value of $u = 500$ for all of the inside points of Fig. 8-4, the matrix values of u at iteration zero will be

Row
↓   0   1   2   3   4   5   6 ← Column

0	500	1,000	1,000	1,000	1,000	1,000	500
1	0	500	500	500	500	500	0
2	0	500	500	500	500	500	0
3	0	500	500	500	500	500	0
4	0	500	500	500	500	500	0
5	0	500	500	500	500	500	0
6	500	1,000	1,000	1,000	1,000	1,000	500

The first iteration starts by averaging the four values around each of the inside points and replacing the old value with this average. The averaging starts with element $(1, 1)$, $(1, 2)$, . . . , $(1, 5)$ and proceeds with $(2, 1)$, $(2, 2)$, . . . , and so on. Thus at the end of the first iteration the matrix of u values will be

ITERATION 1

500	1000	1000	1000	1000	1000	500
0	500	625	656.25	664.063	541.016	0
0	375	500	539.063	550.781	397.949	0
0	343.75	460.938	500	512.695	352.661	0
0	335.938	449.219	487.305	500	338.165	0
0	458.984	602.051	647.339	661.835	500	0
500	1000	1000	1000	1000	1000	500

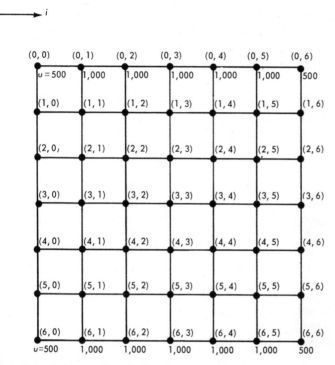

Figure 8-4. *Solution of Laplace's equation by iteration.*

Iterations can proceed in a similar manner resulting in the following values of u for the 10th and 20th iterations

ITERATION 10

500	1000	1000	1000	1000	1000	500
0	500.	682.775	730.923	682.836	500.072	0
0	317.225	500.	557.798	500.125	317.375	0
0	269.077	442.202	500.	442.357	269.264	0
0	317.164	499.875	557.643	500.	317.316	0
0	499.928	682.625	730.736	682.684	500.	0
500	1000	1000	1000	1000	1000	500

ITERATION 20

500	1000	1000	1000	1000	1000	500
0	500.	682.692	730.769	682.692	500.	0
0	317.308	500.	557.692	500.	317.308	0
0	269.231	442.308	500.	442.308	269.231	0
0	317.308	500.	557.692	500.	317.308	0
0	500.	682.692	730.769	682.692	500.	0
500	1000	1000	1000	1000	1000	500

Note that the values shown at iteration 20 are completely symmetrical and agree with the values of Eq. (8.11) to three significant figures.

The following digital computer program was written in BASIC programming language [3] to accomplish the iteration procedure discussed above.

Statement 5 of the program reserves space for the matrix u. Statement 10 sets each element of matrix u equal to unity. Statement 15 multiplies all of the elements of matrix u by 500, which is the initial value of u at iteration zero.

```
5     DIM U(6,6)
10    MAT U=CON
15    MAT U=(500)*U
25    FOR J=1 TO 5
30    LET U(0,J)=1000
32    LET U(6,J)=1000
35    NEXT J
40    LET U(0,0)=500
42    LET U(0,6)=500
44    LET U(6,0)=500
46    LET U(6,6)=500
50    LET Z=0
70    FOR I=1 TO 5
75    LET U(I,0)=0
80    LET U(I,6)=0
85    NEXT I
87    PRINT "ITERATION"Z
88    PRINT
90    MAT PRINT U;
92    LET Z=Z+1
95    FOR I=1 TO 5
100   FOR J=1 TO 5
105   LET U(I,J)=U(I+1,J)+U(I-1,J)+U(I,J+1)+U(I,J-1)
107   LET U(I,J)=U(I,J)/4
110   NEXT J
115   NEXT I
116   PRINT
117   PRINT "ITERATION"Z
120   MAT PRINT U;
130   IF Z<10 THEN 92
135   END
```

Statements 25 through 85 enter the values of u at the four boundaries of Fig. 8-4. (Note that Z in statement 50 is the number of the iteration.)

tatements 87 through 90 print the matrix of u values at zeroth iteration.
tatements 95 through 115 compute values of u using Eq. (8.12). State-
ients 116 through 120 print out the answers. Statement 130 sets an
pper limit to the number of iterations. (Note that statement 130 was
hanged to $Z < 20$ to obtain iteration 20 given above.)

¡ENERAL FORM OF LAPLACE'S EQUATION

The above formulations are specialized forms of a more general
elation describing the finite-difference representation of Laplace's equa-
ion. In a number of more complicated applications the formulas
escribed in the following pages are preferred to the ones given pre-
iously as they enjoy a faster rate of convergence to a solution.

The general form of Laplace's equation can be written in two-
imensional form as

$$\alpha \frac{\partial^2 u}{\partial x^2} + \beta \frac{\partial^2 u}{\partial y^2} = f(x, y) \tag{8.13}$$

vhere α and β may be functions of variables x and y, with the condition
•f elliptic equations satisfied. The above equation is known as Poisson's
quation and specializes to Laplace's equation for $\alpha = \beta = $ constant and
$(x, y) = 0$. The finite-difference representation of Eq. (8.13) for the
qual increments $\Delta x = \Delta y = h$ can be written

$$\gamma_1 u(x + h, y) + \gamma_2 u(x - h, y) + \gamma_3 u(x, y + h) + \gamma_4 u(x, y - h)$$
$$+ \gamma_5 u(x, y) = g(x, y) \tag{8.14}$$

vhere γ's are functions of x, y, and the increment h. Solving the above
quation for $u(x, y)$ and denoting the resulting coefficients by ν's, we
iave

$$u(x, y) = \nu_1 u(x + h, y) + \nu_2 u(x - h, y) + \nu_3 u(x, y + h)$$
$$+ \nu_4 u(x, y - h) + \tau(x, y) \tag{8.15}$$

In addition to the previously discussed solutions, several iterative proce
dures are available for the solution of the above equation. Assuming tha
the values of the variable $u(x, y)$ are available at iteration n over the x,
region and denoting these values by $u_n(x, y)$, the following iteratio
procedures can be formulated.

1. Gauss-Seidel method or the method of *successive displacement*.
The iteration formula for this method is

$$u_{n+1}(x, y) = \nu_1 u_n(x + h, y) + \nu_2 u_{n+1}(x - h, y) + \nu_3 u_n(x, y + h)$$
$$+ \nu_4 u_{n+1}(x, y - h) + \tau_n(x, y) \tag{8.16}$$

In a given problem the initial values of the function $u_0(x, y)$ are give
along the boundaries, as was the case of Fig. 8-4. Assuming a set o
values for inside points at iteration zero, Eq. (8.16) can be solve
successively for points $(1, 1), (1, 2), (1, 3), \ldots, (2, 1), \ldots$ and so on, a
each point using new values of u as they become available. After gettin
to point $(5, 5)$ of the figure the second iteration is started from $(1, 1)$
Continuing this procedure the final solution of the problem is obtained
Note that Eq. (8.16) in the case of Laplace's equation reduces to th
previously discussed Liebmann method.

Equation (8.16) can be slightly modified to improve the rate o
convergence of the solution. This modification is as follows:

$$u_{n+1}(x, y) = \omega[\nu_1 u_n(x + h, y) + \nu_2 u_{n+1}(x - h, y) + \nu_3 u_n(x, y +$$
$$+ \nu_4 u_{n+1}(x, y - h) + \tau_n(x, y)] - (\omega - 1) u_n(x, y) \tag{8.17}$$

The parameter ω is known as the *relaxation factor* and the method
which use this equation are termed as *successive-overrelaxation* methods
Note that for $\omega = 1$ Eq. (8.17) reduces to Eq. (8.16). The choice of ϵ
determines the rate of convergence of the solution. No general expressio
for optimum ω (fastest rate of convergence to a solution) is available fo
use in Eq. (8.17). However, in the case of Laplace's equation $\tau = 0$ an
$\nu_1 = \nu_2 = \nu_3 = \nu_4 = 1/4$, Eq. (8.17) reduces to

$$u_{n+1}(i, j) = \omega\left\{\frac{1}{4}\left[u_n(i+1, j) + u_{n+1}(i-1, j) + u_n(i, j+1)\right.\right.$$
$$\left.\left. + u_{n+1}(i, j-1)\right]\right\} - (\omega - 1)u_n(i, j) \tag{8.18}$$

here i and j are indices in the x and y directions.

The optimum value of ω for this case is given [1] as

$$\omega = 1 + \left(\frac{\bar{\mu}}{1 + \sqrt{1 - \bar{\mu}^2}}\right)^2 \tag{8.19}$$

where

$$\bar{\mu} = \frac{1}{2}\left(\cos\frac{\pi h}{a} + \cos\frac{\pi h}{b}\right)$$

nd a and b are the sides of the rectangle. The above expression for $= b = 1$ and small h becomes

$$\bar{\mu} = \cos\pi h \sim 1 - \frac{\pi^2 h^2}{2}$$

$$\omega = 1 + \left(\frac{\cos\pi h}{1 + \sin\pi h}\right) \sim \frac{2}{1 + \pi h} \tag{8.20}$$

Although the above value of ω applies to the specialized case of Eq. 8.18), it is used as a guide in selecting values of ω in the solution of the more general Eq. (8.17). It can be shown with the ω of Eq. (8.20) that he successive-overrelaxation method converges $2/\pi h$ times faster than he Gauss-Seidel Method.

2. Alternating-direction implicit method. This method, developed by Peaceman and Rachford [1], is along the same lines as the relaxation methods discussed above. The application of this method, however, involves solution of tridiagonal linear simultaneous equations at each tage of iteration. For this reason before discussing the method we will describe the solution of this type of simultaneous equations.

Consider the set of tridiagonal simultaneous equations

$$B_1 T_1 + C_1 T_2 = D_1$$
$$A_i T_{i-1} + B_i T_i + C_i T_{i+1} = D_i \quad (i = 2, 3, \ldots, m-1) \quad (8.21$$
$$A_m T_{m-1} + B_m T_m = D_m$$

where A, B, C, and D's are constants and T's are the variables. The solu tion T's can be expressed

$$T_m = q_m$$
$$T_i = q_i - b_i T_{i+1} \quad (i = m-1, m-2, \ldots, 1) \quad (8.22$$

where

$$q_1 = \frac{D_1}{B_1}, \quad q_i = \frac{D_i - A_i q_{i-1}}{B_i - A_i b_{i-1}}$$
$$\quad (i = 2, 3, 4, \ldots, m-1)$$
$$b_1 = \frac{C_1}{B_1}, \quad b_i = \frac{C_i}{B_i - A_i b_{i-1}}$$
$$(8.23$$

The above method, which can be derived from gaussian elimination method of solving simultaneous equations, is due to Thomas [4]. Tri diagonal simultaneous equations (8.21) arise quite often in numerica solutions of partial and ordinary differential equations.

Getting back to the alternating-direction method of Peaceman and Rachford, the basic iteration equations of this method are, for $\Delta x = \Delta y = h$,

$$u_{n+1/2}(x,y) = u_n(x,y) + r[u(x+h, y) + u(x-h, y) - 2u(x,y)]_{n+1/2}$$
$$+ r[u(x, y+h) + u(x, y-h) - 2u(x,y)]_n$$
$$u_{n+1}(x,y) = u_{n+1/2}(x,y) + r[u(x+h, y) + u(x-h, y) - 2u(x,y)]_{n+1/2}$$
$$+ r[u(x, y+h) + u(x, y-h) - 2u(x,y)]_{n+1}$$
$$(8.24$$

where the subscripts refer to iteration number associated with the variable u and r is a constant. Assuming that the values of u at iteration n $u_n(x, y)$ are available, each complete iteration will require the solution of the first and then the second of Eqs. (8.24). Mechanization of the first of the above equations will result in m tridiagonal simultaneous equations, where m is the number of points inside the rectangle. For example, in the case of the problem of Fig. 8-4, the first of Eqs. (8.24) will give

$$u_{n+1/2}(1,1) = u_n(1,1) + r[u(1,2) + u(1,0) - 2u(1,1)]_{n+1/2}$$
$$+ r[u(2,1) + u(0,1) - 2u(1,1)]_n$$

$$u_{n+1/2}(1,2) = u_n(1,2) + r[u(1,3) + u(1,1) - 2u(1,2)]_{n+1/2} \qquad (8.25)$$
$$+ r[u(2,2) + u(0,2) - 2u(1,2)]_n$$

$$\cdots\cdots\cdots\cdots\cdots\cdots\cdots\cdots\cdots\cdots\cdots\cdots\cdots\cdots\cdots\cdots$$

$$u_{n+1/2}(5,5) = u_n(5,5) + r[u(5,6) + u(5,4) - 2u(5,5)]_{n+1/2}$$
$$+ r[u(6,5) + u(4,5) - 2u(5,5)]_n$$

Transferring the unknowns of the above equations to one side, we get

$$B_1 u_{n+1/2}(1,1) + C_1 u_{n+1/2}(1,2) = D_1$$

$$A_2 u_{n+1/2}(1,1) + B_2 u_{n+1/2}(1,2) + C_2 u_{n+1/2}(1,3) = D_2 \qquad (8.26)$$

$$\cdots\cdots\cdots\cdots\cdots\cdots\cdots\cdots\cdots\cdots\cdots\cdots\cdots\cdots\cdots\cdots$$

$$A_{25} u_{n+1}(5,4) + B_{25} u_{n+1}(5,5) = D_{25}$$

The solutions of the above equations by the procedure outlined in Eqs. (8.22) and (8.23) will result in values of u at iteration $n + 1/2$. In Eqs. (8.26), A, B, C, and D are known quantities which are functions of previously available $u_n(x, y)$. After calculating the values of $u_{n+1/2}$ the second set of Eqs. (8.24) are used in much the same manner to compute u_{n+1}. Note that the first set of Eqs. (8.24) represents a row iteration (x direction) and the second set represents column iteration (y direction) and hence the name *alternating direction*. The term *implicit* in the name of the method is used to denote the fact that solutions at iteration $n + 1$ cannot directly be obtained in terms of solutions at iteration n as was the case with overrelaxation methods. On the other hand when solutions at iteration $n + 1$ can be expressed in terms of solutions at n the method of solution belongs to the general category of *explicit* methods.

It is suggested [1] to use the equation

$$r_k = \frac{1}{(bx^k - 1)} \qquad k = 1, 2, 3, \ldots, t$$

$$x = \left(\frac{a}{b}\right)^{1/(t-1)}$$

(8.27)

for determining the values of the constant r. In the above equation a and b are the sides of the rectangle, k is the number of iterations, and t is the number of the last iteration. Since t is not known at the beginning of a given problem an assumption as to its value is made and the resulting r's are computed. These values are then used in Eqs. (8.24). If the number of iterations t seems unsatisfactory from the convergence viewpoint of resulting solutions, another number is chosen and the calculations repeated.

In general, it can be shown that the number of iterations needed to achieve specific solution accuracies is much less using the alternating-direction method than successive overrelaxation. However, this apparent advantage should be balanced against the difficulty of implementing calculations depicted by Eqs. (8.25) and the selection of parameter r.

SOLUTION OF THE HEAT EQUATION

Consider the one-dimensional heat conduction equation

$$\frac{\partial^2 u}{\partial x^2} = a \frac{\partial u}{\partial t}$$

(8.28)

where $u(x, t)$ is the temperature distribution, x is the space dimension, and t is the time. The constant a depends on the heat transfer properties of the solid in question. For simplicity, assume that $a = 1$. Using the relations

$$\frac{\partial^2 u}{\partial x^2} = \frac{u_{i+1,j} - 2u_{i,j} + u_{i-1,j}}{h^2}$$

$$\frac{\partial u}{\partial t} = \frac{u_{i,j+1} - u_{i,j}}{T}$$

where h and T are the space and time increments Δx and Δt, respectively, Eq. (8.28) becomes

$$u_{i,j+1} = \alpha u_{i+1,j} + (1 - 2\alpha) u_{i,j} + \alpha u_{i-1,j} \tag{8.29}$$

Indices i and j are used for space and time dimensions, respectively. α in Eq. (8.29) is related to h and T by

$$\alpha = \frac{T}{h^2} \tag{8.30}$$

Having the values of the temperature u at time jT, $u(i, j)$ for all i's, the values of the temperature at time $(j + 1)T$, $u(i, j + 1)$, can be obtained using Eq. (8.29). This equation takes an especially simple form when $\alpha = 1/2$, that is,

$$u_{i,j+1} = \frac{1}{2}(u_{i+1,j} + u_{i-1,j}) \tag{8.31}$$

Or, the temperature at a point i at time $j + 1$ is equal to the average of the temperatures at either sides of i, $i + 1$, and $i - 1$, at time j. It can also be shown that the solution represented by Eq. (8.29) is stable only for values of $\alpha \leq 1/2$ (see [2] and the following general discussion). The numerical solution of the heat equation (8.28), using the above method, is illustrated by the following example.

Example. Consider the solution of the temperature distribution equation

$$\frac{\partial^2 u(x, t)}{\partial x^2} = \frac{\partial u(x, t)}{\partial t} \tag{8.32}$$

with the following initial and boundary conditions:

$$u(0, t) = 100 \qquad \text{constant temperature boundary at } x = 0$$

$$\frac{\partial u}{\partial x}(1, t) = 0 \qquad \text{insulated boundary at } x = 1$$

$$u(x, 0) = 0 \qquad \text{initial temperature equal to zero}$$

In order to use matrix notation, the increments and indices are set up as in Fig. 8-5. The insulated boundary condition

$$\frac{\partial u(nh, jT)}{\partial x} = 0 \qquad \text{all } j \tag{8.33}$$

where T is the time increment, can be represented by using backward differences on the space coordinate x at $x = nh$

$$\frac{u_{n,j} - u_{n-1,j}}{\Delta x} = 0 \tag{8.34}$$

or

$$u_{n,j} = u_{n-1,j}$$

That is, when the temperatures of the nth and the $(n - 1)$th rows of Fig. 8-5 are equal, there will be no flow of heat between the regions $(n - 1) h < x \le nh$.

Assuming that the region between $x = 0$ to $x = 1$ is divided into ten equal increments of $h = 0.1$, the value of $n = 1/0.1 = 10$. The initial and boundary conditions now become

$$\begin{aligned} u_{i,0} &= 0 & i &= 1, 2, \ldots, 10 \\ u_{0,j} &= 100 & j &= 1, 2, 3, \ldots \\ u_{n,j} &= u_{n-1,j} & j &= 1, 2, 3, \ldots \end{aligned} \tag{8.35}$$

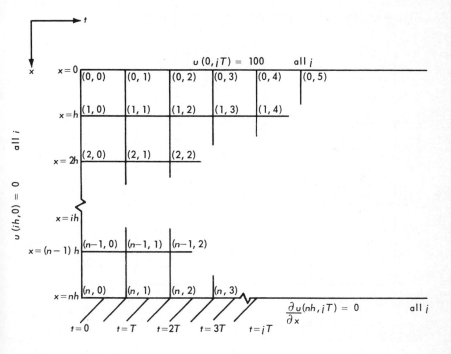

figure 8-5. *Solution of heat flow equation (8.32).*

Having these values, Eq. (8.29) can be solved for $u_{i,j+1}$ for $j = 0$ and $i = 1, 2, 3, 4, \ldots , 9$. The value of $u_{10,0}$ is set equal to $u_{9,0}$ by the last of Eqs. (8.35). Having these values, Eq. (8.29) can now be used to evaluate u's at the next time step by setting $j = 1$ and $i = 1, 2, 3, \ldots , 9$. The value of $u_{10,1} = u_{9,1}$. This procedure can be continued for $j = 2, 3, \ldots ,$ and so on.

Since the value of h was set at 0.1, a value of $T = 0.005$ will result in

$$\alpha = \frac{T}{h^2} = \frac{0.005}{0.01} = 0.5$$

Using this value of α, solutions can be evaluated for $t = 0.005, 0.01, 0.015,$ and so on. These solutions were evaluated by the digital computer program given below, and the results are presented in Fig. 8-6. It is noted that as time increases, the solution approaches the steady-state

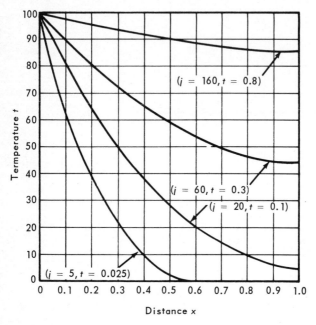

figure 8-6. *Solution of heat equation.*

value of 100.   The digital computer program was written in BASIC
programming language [3] to accomplish the above numerical procedure.

Statement 10 reserves memory spaces for u.   This statement assumes
that there are ten divisions in the x direction and that the time extends
for 160 increments.   Statement 11 sets the value of the index M (time
index) equal to 160.   Statement 15 sets matrix U equal to zero.   Statement 20 sets the value of $A = \alpha = 0.5$ for an $h = 0.1$ and $T = 0.005$.
Statement 25 sets the value of index $N = 1/h = 10$.   Statements 40
through 85 implement the first two equations of (8.35).   Statements 90
through 120 carry out the numerical solution using Eq. (8.29) and the

```
10    DIM U(10,160)
11    LET M=160
15    MAT U=ZER
20    LET A=.5
25    LET N=1/.1
40    FOR I=1 TO N
50    LET U(I,0)=0
60    NEXT I
```

```
70    FOR J=0 TO M
80    LET U(0,J)=100
85    NEXT J
90    FOR J=0 TO M-1
94    LET K=J+1
95    FOR I=1 TO N-1
100   LET U(I,K)=A*U(I+1,J)+U(I,J)*(1-2*A)+A*U(I-1,J)
115   NEXT I
116   LET U(N,K)=U(N-1,K)
120   NEXT J
132   FOR J=0 TO M STEP 20
133   PRINT "J="J
134   FOR I=0 TO M
135   PRINT U(I,J);
136   NEXT I
137   PRINT
138   NEXT J
140   END
```

last of Eqs. (8.35). Statements 132 through 138 print the answers which are the elements of matrix U. These answers are printed for $J = 0$ ($t = 0$), $J = 20$ ($t = 0.005 \times 20 = 0.1$), $J = 40$ ($t = 0.2$), . . . , $J = 160$ ($t = 0.8$) starting with $i = 0, 1, 2, \ldots, 10$, as shown in Table 8-2. Note that the first value of U for each J is equal to 100, which is the boundary condition at $x = 0$.

GENERAL FORM OF THE HEAT EQUATION

The general form of the heat equation in one space dimension can be written as

$$\frac{\partial u}{\partial t} - \sigma \frac{\partial^2 u}{\partial x^2} = 0 \qquad \sigma = \text{const} > 0 \tag{8.36}$$

The boundary conditions $u(x, t)$ and the initial condition $u(x, 0)$, are assumed to be available. This equation can be written in the following finite-difference form:

$$\frac{u_i^{n+1} - u_i^n}{\Delta t} - \sigma \frac{\theta(\delta^2 u)_i^{n+1} + (1 - \theta)(\delta^2 u)_i^n}{(\Delta x)^2} = 0 \tag{8.37}$$

Table 8-2. Results of the Digital Computer Program for the Heat Equation

J										
J=0	100	0	0	0	0	0	0	0	0	0
J=20	100	82.3823	66.3664	50.3666	38.3713	26.4666	19.1824	12.2516	9.01718	6.79932
J=40	100	87.7503	76.0976	64.6905	54.9116	45.7011	38.7324	32.7746	29.1705	27.1129
J=60	100	90.7983	81.9273	73.4053	65.8137	58.9255	53.3994	48.9253	46.0197	44.4693
J=80	100	93.023	86.262	79.8276	74.0097	68.8246	64.5687	61.2071	58.9616	57.7978
J=100	100	94.7019	89.5567	84.6806	80.2439	76.3212	73.0701	70.5299	68.8126	67.9337
J=120	100	95.9753	92.0631	88.3622	84.9859	82.0109	79.535	77.5096	76.3012	75.6352
J=140	100	96.9422	93.9687	91.158	88.5908	86.3321	84.4491	82.9877	81.9923	81.4869
J=160	100	97.6767	95.417	93.2818	91.3306	89.615	88.1835	87.0736	86.317	85.9331

where i is the space and n is the time index $(x = i\,\Delta x, t = n\,\Delta t)$ and the δ^2 operator is used to designate

$$(\delta^2 u)_m^{k} = u_{m+1}^k - 2u_m^{k} + u_{m-1}^k \tag{8.38}$$

and θ is a positive constant between zero and one. Thus, Eq. (8.37) in the case of $\theta = 0$ specializes to

$$u_i^{n+1} - u_i^{n} - \sigma\left(\frac{\Delta t}{\Delta x^2}\right)\left(u_{i+1}^n - 2u_i^{n} + u_{i+1}^n\right) = 0 \tag{8.39}$$

Using the notation

$$\alpha = \sigma\,\frac{\Delta t}{\Delta x^2}$$

Eq. (8.39) becomes

$$u_i^{n+1} = \alpha u_{i+1}^n + (1 - 2\alpha)\,u_i^{n} + \alpha u_{i-1}^n \tag{8.40}$$

This equation, but for indexing, represents Eq. (8.29). For any other value of θ $(\theta \neq 0)$ Eq. (8.37) will result in a set of simultaneous equations to be solved at each iteration. The condition for the stability of solution of Eq. (8.37) is, from [2],

$$\frac{2\sigma\,\Delta t}{(\Delta x)^2} \leq \frac{1}{1 - 2\theta} \qquad \text{if } 0 \leq \theta < \tfrac{1}{2} \tag{8.41}$$

$$\text{No restriction} \qquad \text{if } \tfrac{1}{2} \leq \theta \leq 1$$

It can be further shown that the error in representation (8.39) is of the order

$$e(u) = 0(\Delta t) + 0[(\Delta x)^2] \tag{8.42}$$

By varying the values of θ in Eq. (8.37) a number of relations can be obtained for solutions $u(x, t)$ (see [2]). One such solution is for $\theta = 1/2$

$$\frac{u_i^{n+1} - u_i^n}{\Delta t} = \sigma \frac{(\delta^2 u)_i^n + (\delta^2 u)_i^{n+1}}{2\Delta x^2} \tag{8.43}$$

developed by Crank and Nicholson in 1947. It can be shown that this solution is always stable ($0 \leq \theta \leq 1$) and the truncation error is of the order of magnitude

$$e = 0[(\Delta t)^2] + 0[(\Delta x)^2] \tag{8.44}$$

Expansion of Eq. (8.43) will result in

$$u_i^{n+1} - u_i^n = \frac{\sigma \Delta t}{2\Delta x^2}\Big([u_{i+1} - 2u_i + u_{i-1}]^n$$
$$+ [u_{i+1} - 2u_i + u_{i-1}]^{n+1}\Big) \tag{8.45}$$

Noting that the values of u at time step n are available, the above equation can be put in the form

$$A_i u_{i-1}^{n+1} + B_i u_i^{n+1} + C_i u_{i-1}^{n+1} = D_i \qquad i = 2, 3, \ldots, M-1 \tag{8.46}$$

where M is the number of points in the x direction. And, since the boundary conditions are known, the equations representing the boundaries will be of the form

$$B_1 u_1^{n+1} + C_1 u_2^{n+1} = D_1$$
$$A_M u_{M-1}^{n+1} + B_M u_{M-1}^{n+1} = D_M \tag{8.47}$$

The solution of Eqs. (8.46) and (8.47) can be obtained by the method of Eq. (8.21), at each iteration. Note that in formulations where $\theta \neq 0$ Eq. (8.37) results in a set of simultaneous equations to be solved at each iteration.

Formulas similar to Eq. (8.37) can be derived for the solution of the two-dimensional heat equation

$$\frac{\partial u}{\partial t} = A \frac{\partial^2 u}{\partial x^2} + C \frac{\partial^2 u}{\partial y^2} \tag{8.48}$$

where A and C may be functions of x and y. Denoting

$$u_{i,j}^n = u(i \, \Delta x, \, j \, \Delta y, \, n \, \Delta t) \tag{8.49}$$

and

$$\varphi_{i,j}^n = \frac{A}{\Delta x^2} \left(u_{i+1,j}^n - 2u_{i,j}^n + u_{i-1,j}^n \right) + \frac{C}{\Delta y^2} \left(u_{i,j+1}^n - 2u_{i,j}^n + u_{i,j-1}^n \right) \tag{8.50}$$

We can write the finite difference representation of Eq. (8.48) by

$$\frac{u_{i,j}^{n+1} - u_{i,j}^n}{\Delta t} = \theta \varphi_{i,j}^{n+1} + (1 - \theta) \varphi_{i,j}^n \tag{8.51}$$

where again θ represents a number between zero and one. For $\theta > 0$ the above equations result in a system of simultaneous equations to be solved at each iteration. For $\theta = 0$, of course, this is not required as the equations for u^{n+1} will be in terms of u^n known quantities. The conditions of stability of solution of Eq. (8.51) are

$$\frac{A \, \Delta t}{(\Delta x)^2} + \frac{B \, \Delta t}{(\Delta y)^2} \leq \frac{1}{2 - 4\theta} \qquad \text{if } 0 \leq \theta < \frac{1}{2}$$

$$\text{No restriction} \qquad \qquad \text{if } \frac{1}{2} \leq \theta \leq 1$$

This can be considered as generalization of stability condition found in the one-dimensional case.

SOLUTION OF THE WAVE EQUATION

The wave equation, which represents the position of a vibrating string as a function of time (Fig. 8-7), can be represented as

$$c^2 \frac{\partial^2 y}{\partial x^2} = \frac{\partial^2 y}{\partial t^2} \tag{8.52}$$

where $y(x, t)$ is the vertical position at distance x and time t and c^2 is a constant depending on the physical properties of the string.

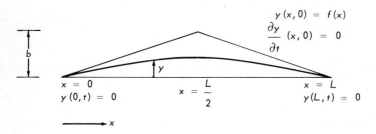

figure 8-7.   *Coordinates of wave equation.*

Assume the following initial conditions for the wave equation (8.52)

$$y(x, 0) = f(x)$$
$$\frac{\partial y}{\partial t}(x, 0) = 0 \tag{8.53}$$

These equations give displacement and velocity of the string at time zero. The assumed boundary conditions at $x = 0$ and $x = L$, where L is the length of the string, are as follows:

$$y(0, t) = 0$$
$$y(L, t) = 0 \tag{8.54}$$

These equations state that the end points of the string are fixed. Let us denote the deflection at x_i at time t_j by $y_{i,j}$, and the space and time ncrements by h and T, respectively. Having done this, Eq. (8.52) can be written in central difference form (of order h^2) as follows:

$$y_{i+1,j} - 2y_{i,j} + y_{i-1,j} = \frac{1}{\alpha^2}(y_{i,j+1} - 2y_{i,j} + y_{i,j-1}) \qquad (8.55)$$

where

$$\alpha^2 = \frac{c^2 T^2}{h^2}$$

The value of $y_{i,j+1}$ can be obtained in terms of values of y at the previous time steps from Eq. (8.55). This results in

$$y_{i,j+1} = \alpha^2(y_{i+1,j} + y_{i-1,j}) + 2(1 - \alpha^2)y_{i,j} - y_{i,j-1} \qquad (8.56)$$

For $\alpha = 1$ (in [2] it is shown that solutions will be stable only if $\alpha \leq 1$), this equation reduces to the simple form

$$y_{i,j+1} = (y_{i+1,j} + y_{i-1,j}) - y_{i,j-1} \qquad (8.57)$$

Let us assume that the length of the string is divided into n parts where $n = L/h$. The initial and boundary conditions (8.53) and (8.54) can now be written

$$y_{i,0} = f(x_i) \qquad\qquad i = 0 \text{ to } n$$

$$\frac{y_{i,-1} - y_{i,1}}{2T} = 0, \quad y_{i,-1} = y_{i,1}, \quad i = 0 \text{ to } n$$

$$y_{0,j} = 0 \qquad\qquad j = 1, 2, 3, \ldots \qquad (8.58)$$

$$y_{n,j} = 0 \qquad\qquad j = 1, 2, 3, \ldots$$

n the second of the above equations, central differences are used to represent the initial zero velocity condition.

Equation (8.56) for any i and $j = 0$ can be written

$$y_{i,1} = \alpha^2(y_{i+1,0} + y_{i-1,0}) + 2(1 - \alpha^2)y_{i,0} - y_{i,-1} \qquad (8.59)$$

The second of Eqs. (8.58) states that $y_{i,-1} = y_{i,1}$; substituting in Eq (8.59), we get

$$y_{i,1} = \frac{\alpha^2}{2}(y_{i+1,0} + y_{i-1,0}) + (1 - \alpha^2)y_{i,0} \qquad (8.60)$$

This equation can be used for obtaining the value of $y_{i,1}$ for $i = 1, 2, 3, \ldots$ $n-1$. Note that the values of $y_{0,1}$ and $y_{n,1}$ are given by the boundary conditions (8.58). Having these values, and setting $j = 1$ in Eq. (8.56) the values of $y_{i,2}$, $i = 1, 2, 3, \ldots, n-1$ can be obtained for $t = 2\Delta t = 2T$ The procedure can be carried out for $j = 3, 4, 5, \ldots$, and so on.

Example. As a numerical example of the above equations, consider the problem of the vibrating string with the initial conditions given in Fig. 8-7. Dividing the length of the string in n parts ($h = L/n$), the initial and boundary conditions (8.58) will become

$$y_{i,0} = b(i)\left(\frac{L}{n}\right) \qquad \text{for } 0 \leq i \leq \frac{n}{2}$$

$$= b(n - i)\left(\frac{L}{n}\right) \qquad \text{for } \frac{n}{2} \leq i \leq n \qquad (8.61)$$

$$y_{i,-1} = y_{i,1}$$

$$y_{0,j} = 0$$

$$y_{n,j} = 0$$

where b is the initial height at the center of the string. Assuming the length of the string $L = 2$, the initial height $b = 1/2$, and dividing the length into 10 equal parts ($n = 10$), we get

$$h = \frac{L}{n} = \frac{2}{10} = 0.2$$

An α and c of unity will result in

$$\alpha^2 = \frac{c^2 T^2}{h^2} = \frac{T^2}{h^2} = 1 \qquad (8.62)$$

or

$$T = 0.2$$

Using Eq. (8.60) for $i = 1$ to $n - 1$ will give the values of $y_{i,1}$ for the time $T = 0.2$. Note that the value of $y_{n,1}$ is zero by the boundary condition. After these values are calculated, Eq. (8.59) can be applied for $j = 1$ and $i = 1, 2, 3, \ldots, n - 1$ to compute the positions of the string for $j = 2$ or time $= t = 2(0.2) = 0.4$. The solution can proceed by calculating $y_{i,j}$ for $j = 3$ and $i = 1, 2, 3, \ldots, n - 1$ and so on for $j = 4, 5, \ldots$.

The analytical solution of the vibrating string problem with the given initial and boundary conditions can be obtained from [5]. This solution is given in an infinite trigonometric series as follows

$$y(x, t) = \frac{8b}{\pi^2} \sum_{m=1}^{\infty} \frac{1}{m^2} \sin \frac{m\pi}{L} \sin \frac{m\pi x}{L} \cos \frac{n\pi at}{L} \tag{8.63}$$

The numerical solution described above and the analytical solution (8.63) were evaluated with the aid of digital computers. The series of Eq. (8.63) was summed from $m = 1$ to 100. Table 8-3 shows the partial results of the calculations. An examination of this table shows that the values calculated by the numerical method are extremely close to the analytical solution. For $\alpha = 1$, it can be shown that the numerical solution is identical to the analytical solution [6]. The slight discrepancy in the values as given in Table 8-3 is primarily due to the approximation in the evaluation on the series (8.63), that is, m to 100 instead of $m = \infty$. The accuracy of the numerical solution decreases slightly as α assumes values smaller than one.

For purposes of comparison, the series of Eq. (8.63) was also evaluated summing up to $m = 5$. The results are presented in Table 8-4 for time equal to 0.6. Note that the finite-difference solution of the table is the exact solution.

SUMMARY

The solutions given above were primarily derived for linear partial differential equations. However, the methods can be extended to nonlinear partial differential equations. One such application is given in [7].

Table 8-3. Results of the Solution of the Wave Equation

DIST X	Time = 0		Time = 0.2	
	Y-ANLYT	Y-FINITE DIFF	Y-ANLYT	Y-FINITE DIFF
0	0	0	0	0
.2	9.99999 E-2	.1	9.99998 E-2	.1
.4	.2	.2	.199999	.2
.6	.299999	.3	.299998	.3
.8	.399996	.4	.398986	.4
1.	.497974	.5	.399996	.4
1.2	.399996	.4	.398986	.4
1.4	.299999	.3	.299998	.3
1.6	.2	.2	.199999	.2
1.8	9.99998 E-2	.1	9.99998 E-2	.1
2.	−1.21982 E-8	0	−1.30857 E-8	0

DIST X	Time = 1.0		Time = 2.0	
	Y-ANLYT	Y-FINITE DIFF	Y-ANLYT	Y-FINITE DIFF
0	0	0	0	0
.2	−5.32606 E-10	−1.04774 E-9	−9.99999 E-2	−.1
.4	−1.84689 E-9	−1.62981 E-9	−.2	−.2
.6	−3.14625 E-9	−2.09548 E-9	−.299999	−.3
.8	−3.82612 E-9	−2.79397 E-9	−.399996	−.4
1.	−5.67606 E-9	−2.32831 E-9	−.497974	−.5
1.2	−6.65077 E-9	−2.79397 E-9	−.399996	−.4
1.4	−8.00339 E-9	−2.09548 E-9	−.299999	−.3
1.6	−9.46699 E-9	−1.62981 E-9	−.2	−.2
1.8	−1.08282 E-8	−1.04774 E-9	−9.99999 E-2	−.1
2.	−2.79613 E-14	0	1.31982 E-8	0

The stability and accuracy of the numerical solutions described above have been the subject of extensive study by mathematicians and engineers. Reference 1 contains several chapters that discuss the solution of partial differential equations.   References 8 and 9 compare numerical and analytical solutions of specific partial differential equations.

PROBLEMS

1. Derive the expressions of Eq. (8.4).
2. Divide Fig. 8-3 into four equal increments in each direction and formulate a table of coefficients corresponding to Table 8-1. Obtain the numerical solution and compare with the values given.

Table 8-4. Comparison of Solutions of the Wave Eq. (8.63)

DIST X	Time = 0.6		
	$m = 5$	$m = 100$	
	Y-ANLYT	Y-ANLYT	Y-FINITE DIFF
0	0	0	0
.2	.108262	.10099	.1
.4	.180754	.189964	.2
.6	.205959	.201149	.2
.8	.201388	.199561	.2
1.	.195393	.200318	.2
1.2	.201388	.199561	.2
1.4	.205959	.201149	.2
1.6	.180754	.189964	.2
1.8	.108262	.10099	.1
2.	-1.47587 E-8	-1.17699 E-8	0

3. Solve Prob. 2 by Liebmann's iteration method.
4. Derive Eq. (8.17) for the case of Laplace's equation (8.8) for $\omega = 1/2$.
5. Derive Eqs. (8.20) from Eq. (8.19).
6. Prove relations (8.22) for the set of equations

$$B_1 T_1 + C_1 T_2 \qquad = D_1$$
$$A_2 T_1 + B_2 T_2 + C_2 T_3 = D_2$$
$$A_3 T_2 + B_3 T_3 = D_3$$

7. Solve the set of equations

$$\begin{bmatrix} -1.5 & 1 & 0 & 0 & 0 \\ 1 & -1.5 & 1 & 0 & 0 \\ 0 & 1 & -1.5 & 1 & 0 \\ 0 & 0 & 1 & -1.5 & 1 \\ 0 & 0 & 0 & 1 & -1.5 \end{bmatrix} \begin{pmatrix} X_1 \\ X_2 \\ X_3 \\ X_4 \\ X_5 \end{pmatrix} = \begin{pmatrix} -1 \\ 0.1 \\ 0.1 \\ 0.1 \\ -1 \end{pmatrix}$$

by the method of Eqs. (8.22) and (8.23).

8.   Solve equations

$$
\begin{bmatrix}
25 & 1 & 0 & 0 \\
1 & 16 & 3 & 0 \\
0 & 1 & -12 & -1 \\
0 & 0 & 1 & 5
\end{bmatrix}
\begin{pmatrix}
X_1 \\
X_2 \\
X_3 \\
X_4
\end{pmatrix}
=
\begin{pmatrix}
27 \\
45 \\
-48 \\
14
\end{pmatrix}
$$

by the method of Eqs. (8.22) and (8.23).

9.   Formulate the problem depicted in the example of Fig. 8-3 by the successive-overrelaxation method of Eq. (8.17) with $\omega = 1.5$.

10.   Formulate the problem depicted in Fig. 8-3 by alternating direction method of Eq. (8.24). Draw the logic block diagram of the solution for a digital computer program.

11.   Formulate the solution of the steady-state temperature distribution problem of the rectangle of Fig. P-11.

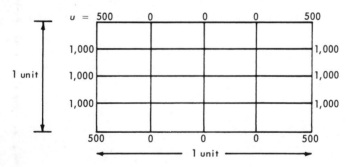

fig. P-11.

The boundary conditions and space increments are specified. Use the alternating-direction method with $r = 0.5$. Obtain numerical values for one complete iteration.

12.   Obtain the values for $J = 5$, $t = 0.025$ of Fig. 8-6 for the given heat transfer problem.

13.   Formulate the heat transfer problem depicted in Fig. 8-5, using the Crank-Nicholson formula (8.43).

14.   Formulate the heat transfer problem of Fig. 8-5, using four increments in the X direction, by the Crank-Nicholson method of Eq. (8.43). Obtain numerical values for the first iteration.

5. For $\theta = 1/2$ formulate equations resulting from the expansion of Eq. (8.51).
6. Obtain the finite-difference solution of the wave equation for $t = 0.2$ as tabulated in Table 8-3.

REFERENCES

. Todd, John (ed.): "Survey of Numerical Analysis," chap. 11, McGraw-Hill Book Company, New York, 1962.

. Richtmyer, R. D., and K. W. Morton: "Difference Methods for Initial Value Problems," Interscience Publishers, Inc., New York, 1967.

. BASIC Language, reference manual, General Electric Company, May, 1966.

. Thomas, L. H.: "Elliptic Problems in Linear Difference Equations," Watson Scientific Laboratory, Columbia University, New York, September, 1949.

. Pipes, L. A.: "Applied Mathematics for Engineers and Physicists," p. 431, McGraw-Hill Book Company, New York, 1958.

. Salvadori, M. G., and M. L. Baron: "Numerical Methods in Engineering," p. 267, Prentice-Hall, Inc., Englewood Cliffs, N. J., 1964.

. Hovanessian, S. A., and F. J. Fayers: Linear Waterflood with Gravity and Capillary Effects, *J. Soc. Petroleum Eng., Trans. AIME*, March, 1961.

. Douglas, J.: "On the Numerical Integration of $\partial^2 u/\partial x^2 + \partial^2 u/\partial y^2 = \partial u/\partial t$ by Implicit Methods," *J. Soc. Ind. Appl. Math.*, 1955, p. 42.

9 LINEAR AND SEPARABLE PROGRAMMING

This discussion is intended to give the reader a basic understanding of linear programming and enable him to recognize and formulate linear programming problems. The solution of these problems can be obtained by the simplex method, which is discussed briefly. For a more complete discussion, the reader is referred to [1]. The simplex method, depending on the size of the problem, involves a great number of arithmetic calculations. Therefore, the method leads naturally to solution by digital computers. Actually the development of digital computers has greatly furthered the utilization of linear programming in the solution of many industrial problems. Today, practically in every industrial or research center having access to a digital computer one can utilize digital computer codes written for the solution of linear programming problems.

The linear programming methods have been applied to the solution of technical as well as economic, scheduling, and allocation problems [1].

The methods have been very successfully applied in the petroleum industry for the solution of gasoline blending problems [1] and economic optimization of oil well development and exploration [2]. The linear programming methods, however, are limited by the fact that, as the name implies, they can be applied only to the solution of problems whose fundamental characteristics are expressed by linear equations. This limitation has been somewhat reduced by the emergence of a number of methods that extend the region of applicability of linear programming to restricted nonlinear regions [3]. The simplest of these methods is the method of *separable programming* [4] which will be discussed in this chapter. This method is an extension of linear programming which allows the inclusion of nonlinear constraints in the formulation of the problem. The method of solution of these problems is identical to the simplex method of linear programming.

FORMULATION AND METHOD OF SOLUTION

Suppose that we have a set of n nonhomogeneous equations in n unknowns, as follows:

$$\sum_{j=1}^{n} a_{ij}x_j = b_i \quad i = 1, 2, \ldots, n \tag{9.1}$$

This set of equations has a unique solution for x_j's, provided that the determinant formed from the a_{ij} coefficients is nonzero. Notice that in the above, the number of equations and the number of unknowns are the same, namely, n. Now, consider the case where there are more equations than unknowns, i.e.,

$$\sum_{j=1}^{m} a_{ij}x_j = b_i \quad i = 1, 2, 3, \ldots, n \tag{9.2}$$

with $n > m$. In this case, as long as the equations of the set are independent, there is no solution at all to the system of equations. Finally, consider the case where there are more unknowns than equations:

$$\sum_{j=1}^{m} a_{ij}x_j = b_i \qquad i = 1, 2, 3, 4, \ldots, n \tag{9.3}$$

with $m > n$. In this case there is no unique solution to the system of Eqs. (9.3). Introducing the additional condition

$$\sum_{j=1}^{m} a_{ij}x_j = \text{minimum} \qquad i = 0 \tag{9.4}$$

and the restriction that the solutions should result in $x_j > 0$ (negative values not accepted) may lead to a unique solution of the linear programming type or, with these restrictions, there may not be a solution to the system at all. In the latter case the linear programming problem is termed *infeasible*. In summary, the set of relations

$$\sum_{j=1}^{m} a_{ij}x_j = b_i \qquad i = 1, 2, 3, \ldots, n \tag{9.5a}$$

$$\sum_{j=1}^{m} a_{ij}x_j = \text{minimum} \qquad i = 0 \tag{9.5b}$$

$$x_j > 0 \qquad j = 1, \ldots, m \tag{9.5c}$$

with $m > n$, constitute a *linear programming* problem. Equations (9.5a) are called the *constrains* or the *restraint equations* while Eq. (9.5b) is the *functional* and is also referred to as the objective form. Equation (9.5c) is the *nonnegativity* requirement on the variables. Observe that the number of nonzero x_j's given by the linear programming solution will be equal to the number of constraints.

To become familar with the terminology and the simplex method of solution of linear programming problems, consider the following numerical example:

$$3x_1 - 2x_2 \leq 12 \tag{9.6a}$$

$$-x_1 + 2x_2 \leq 4 \tag{9.6b}$$

$$x_1 + 2x_2 \geq 6 \tag{9.6c}$$

$$-3x_1 - x_2 = F = \text{minimum} \tag{9.6d}$$

$$x_1, \quad x_2 \geq 0 \tag{9.6e}$$

Equations (9.6a), (9.6b), and (9.6c) are the constraints, while Eq. (9.6d) represents the functional to be minimized. It will be seen later that the inequality constraints (9.6a)-(9.6c) are converted to equality constraints by addition of *slack* variables. Addition of these variables will increase the number of unknowns and render the problem into the form where the number of the unknowns is greater than the number of equations. (If the functional is to be maximized, the coefficients of the variables in the functional are multiplied by -1. Then the minimization of the resulting functional is equivalent to maximizing the original functional.) Equation (9.6e) gives the nonnegativity requirement on the variables x_1 and x_2. Note that all of the constraints have positive right-hand sides. The above constraints and the functional are represented graphically in Fig. 9-1. In this figure, the constraint $3x_1 - 2x_2 \leq 12$ represents the region to the left of the line AA, the constraint $-x_1 + 2x_2 \leq 4$ represents the region below the line BB, and the constraint $x_1 + 2x_2 \geq 6$ represents the region above the line CC. Thus, the region covered by small squares represents the allowable region for the solution of the problem. The dashed lines are for various values of the functional F. From this figure, the constrained minimum of the function F occurs at $x_1 = 8$ and $x_2 = 6$ and has a value of $F = -30$. Now let us solve the same problem by the simplex method.

First, replace the inequality constraints (9.6a), (9.6b), and (9.6c) by equality constraints. This is accomplished by adding slack variables x_3, x_4, x_5, and x_6 to the system, as follows:

$$3x_1 - 2x_2 + x_3 \qquad\qquad\quad = 12 \tag{9.7a}$$

$$-x_1 + 2x_2 \qquad + x_4 \qquad\quad = 4 \tag{9.7b}$$

$$x_1 + 2x_2 \qquad\qquad + x_5 - x_6 = 6 \tag{9.7c}$$

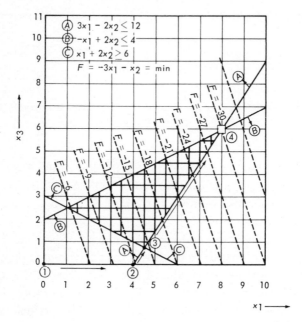

figure 9-1. *Simplex method for the example problem.*

$$-3x_1 - x_2 \qquad\qquad\qquad = F \qquad\qquad (9.7d)$$

$$x_1, \quad x_2, \quad x_3, \quad x_4, \quad x_5, \quad x_6 \ge 0 \qquad\qquad (9.7e)$$

Note that Eqs. (9.6a) and (9.6b) are "smaller than" inequalities and they can be changed to equalities (9.7a) and (9.7b) by adding positive slack variables x_3 and x_4. On the other hand, Eq. (9.6c) is a "greater than" inequality and it can be changed to an equality by adding a slack variable x_6 with a -1 coefficient. But this, as will be seen later, is not computationally desirable. Therefore, add another slack variable x_5 with a $+1$ coefficient. At this point the constraint (9.6c), represented by equality (9.7c), will be satisfied only when $x_5 = 0$ and $x_6 \ge 0$. Thus, as long as $x_5 \ge 0$ we will not be in the allowable region of Fig. 9-1 and the solutions will be somewhat artificial.* Equation (9.7d), as before, represents the functional to be minimized, and Eq. (9.7e) is the nonnegativity requirement on all of the variables.

---

*Slack variables performing functions similar to that of x_5 are sometimes called artificial slacks. In some problems, in order to derive these variables out of the final solution, they are included in the functional with arbitrarily large coefficients (penalty coefficients).

Start the simplex method by taking the solution

$$x_1 = 0 \quad x_2 = 0 \quad x_3 = 12 \quad x_4 = 4 \quad x_5 = 6 \quad x_6 = 0$$

which makes $F = 0$ for the first iteration. Observe that this solution satisfies the constraints ((9.7a), (9.7b), and (9.7c)) and the nonnegativity requirement (9.7e). In linear programming, solutions of this type, having the same number of nonzero variables as there are constraints, are called *basic* solutions, and the nonzero variables are said to be in basis. At this point the question is posed—can the functional assume a lower value? The answer is yes, since, in the functional (9.7d), x_1 and x_2 have negative coefficients. Therefore, increasing the value of either one from zero (given by iteration 1) will decrease the value of the functional. In particular, since x_1 has a larger negative coefficient, a higher rate of decrease of the functional can be obtained by increasing x_1. The amount that x_1 should increase can be obtained from the following observations. Increasing the value of x_1 in a particular constraint will result in an increase, a decrease, or no change in the values of the other variables present in the constraint. We need not concern ourselves with variables which increase or remain unchanged, but should be careful that decreasing variables do not become negative. This might arise only in constraints in which x_1 appears with a positive coefficient. Thus, we only examine constraints (9.7a) and (9.7c) of the problem. From Eq. (9.7a), bringing x_1 in basis will decrease the value of x_3 (which is in basis in iteration 1). The highest value that x_1 can have (with $x_3 = 0$) is $12/3 = 4$. Similarly, from Eq. (9.7c) x_1 can attain a value of 6 with $x_5 = 0$. Thus, take the smaller of these values, 4, and bring x_1 in basis through the constraint (9.7a). Dividing Eq. (9.7a) by the coefficient of x_1, we obtain the set of equations

$$x_1 - \frac{2}{3}x_2 + \frac{1}{3}x_3 \qquad\qquad = 4 \qquad\qquad\qquad (9.8a)$$

$$-x_1 + 2x_2 \qquad + x_4 \qquad = 4 \qquad\qquad\qquad (9.8b)$$

$$x_1 + 2x_2 \qquad\qquad + x_5 - x_6 = 6 \qquad\qquad\qquad (9.8c)$$

$$-3x_1 - x_2 \qquad\qquad\qquad = F \qquad\qquad\qquad (9.8d)$$

$$x_1, \quad x_2, \quad x_3, \quad x_4, \quad x_5, \quad x_6 \geq 0 \qquad\qquad (9.8e)$$

Eliminating x_1 from Eqs. (9.8b), (9.8c), and (9.8d) by multiplying Eq (9.8a) by proper constants and adding to Eqs. (9.8b), (9.8c), and (9.8d) results in

$$x_1 - \frac{2}{3}x_2 + \frac{1}{3}x_3 \qquad\qquad = 4 \qquad\qquad (9.9a)$$

$$\frac{4}{3}x_2 + \frac{1}{3}x_3 + x_4 \qquad = 8 \qquad\qquad (9.9b)$$

$$\frac{8}{3}x_2 - \frac{1}{3}x_3 \qquad + x_5 - x_6 = 2 \qquad\qquad (9.9c)$$

$$-3x_2 + \quad x_3 \qquad\qquad = F + 12 \qquad (9.9d)$$

The basic solution of this iteration is

$$x_1 = 4 \quad x_2 = 0 \quad x_3 = 0 \quad x_4 = 8 \quad x_5 = 2 \quad x_6 = 0$$

This results in

$$F + 12 = 0$$

or

$$F = -12$$

Thus, in this iteration the value of the functional F is decreased by 12 units. Further decrease of the functional can be obtained by increasing x_2, which has a negative coefficient in Eq. (9.9d). Consider constraints (9.9b) and (9.9c), in which x_2 appears with a positive coefficient. From Eq. (9.9b) we note that x_2 can attain a limiting value of $(8)/(4/3) = 6$ and from Eq. (9.9c) a limiting value of $(2)/(8/3) = 3/4$. Thus, we bring x_2 in basis through constraint (9.9c). Dividing Eq. (9.9c) by 8/3, we have

$$x_1 - \frac{2}{3}x_2 + \frac{1}{3}x_3 \qquad\qquad = 4 \qquad\qquad (9.10a)$$

$$\frac{4}{3}x_2 + \frac{1}{3}x_3 + x_4 \qquad = 8 \qquad\qquad (9.10b)$$

$$x_2 - \frac{1}{8}x_3 \qquad + \frac{3}{8}x_5 - \frac{3}{8}x_6 = \frac{3}{4} \qquad (9.10c)$$

$$-3x_2 + \quad x_3 \qquad\qquad = F + 12 \qquad (9.10d)$$

Eliminating x_2 from Eqs. (9.10a), (9.10b), and (9.10d), we get

$$x_1 + \frac{1}{4}x_3 + \frac{1}{4}x_5 - \frac{1}{4}x_6 = \frac{9}{2} \tag{9.11a}$$

$$\frac{1}{2}x_3 + x_4 - \frac{1}{2}x_5 + \frac{1}{2}x_6 = 7 \tag{9.11b}$$

$$x_2 - \frac{1}{8}x_3 + \frac{3}{8}x_5 - \frac{3}{8}x_6 = \frac{3}{4} \tag{9.11c}$$

$$\frac{5}{8}x_3 + \frac{9}{8}x_5 - \frac{9}{8}x_6 = F + \frac{57}{4} \tag{9.11d}$$

The basic solution of the above iteration is

$$x_1 = \frac{9}{2} \quad x_2 = \frac{3}{4} \quad x_3 = 0 \quad x_4 = 7 \quad x_5 = 0 \quad x_6 = 0$$

This solution gives

$$F = -\frac{57}{4} = -14.25$$

Observe that in this iteration x_5 is zero and the solution is no longer artificial. We refer to this solution as a basic *feasible* solution. From Eq. (9.11d) we observe that the functional F can be further decreased by increasing x_6, which appears with a negative coefficient. Equation (9.11b) gives the largest value that x_6 can have, that is, $(7)/(1/2) = 14$. Thus we bring x_6 in basis through Eq. (9.11b) by eliminating x_6 in Eqs. (9.11a), (9.11c), and (9.11d). This results:

$$x_1 + \frac{1}{2}x_3 + \frac{1}{4}x_4 = 8 \tag{9.12a}$$

$$x_3 + 2x_4 - x_5 + x_6 = 14 \tag{9.12b}$$

$$x_2 + \frac{1}{4}x_3 + \frac{3}{8}x_4 = 6 \tag{9.12c}$$

$$\frac{7}{4}x_3 + \frac{9}{4}x_4 = F + 30 \tag{9.12d}$$

The basic feasible solution of this iteration is

$$x_1 = 8 \quad x_2 = 6 \quad x_3 = 0 \quad x_4 = 0 \quad x_5 = 0 \quad x_6 = 14$$

This solution results in a functional $F = -30$. Since negative coefficients do not appear in Eq. (9.12d), the basic feasible solution

$$x_1 = 8 \quad x_2 = 6 \quad \text{and} \quad x_6 = 14$$

with a functional of $F = -30$ is the optimal solution.

The iterations together with the corresponding functional values are expressed in Fig. 9-1. The first iteration with $x_1 = 0$ and $x_2 = 0$, $F = 0$ is at point (1); the second iteration with $x_1 = 4$, $x_2 = 0$, $F = -12$ is at point (2); and so on. Observe that each iteration moves the solution from one corner to the other, while decreasing the value of the functional. Note that, since at iteration 3 the solution becomes feasible, point (3) is a point of the allowable region.

The simplex method, as described above, and its variations are used for the solution of linear programming problems. The simplex method can easily be coded for digital computers. The rules to be followed in the solution of linear programming problems by the simplex method can be summarized as follows:

1. Find the x_i with the highest negative coefficient in the functional.
2. Locate the constraints in which x_i has positive coefficients.
3. Divide the right-hand side of these constraints by the respective x_i coefficients.
4. Choose the smallest of the above values and the respective constraints.
5. Divide this constraint by the coefficient of x_i and eliminate x_i from all of the other constraints and the functional.
6. Repeat the above procedure.

The discussion of the linear programming method has thus far covered only the basic principles of the method. The theory and the computational aspects of linear programming have developed considerably beyond the presentation of this chapter. The interested reader is referred to [5, 6] for further study of the subject.

SEPARABLE PROGRAMMING

Separable programming is a generalization of the simplex method of linear programming for certain types of nonlinear problems. This generalization of the method was developed by C. E. Miller and presented t the 1960 National Meeting of the Association for Computing Machinery [4]. A brief description of the method and its extension to he three-dimensional case is given.

Consider the nonlinear curve $y = f(x)$ of Fig. 9-2 and assume that it an be replaced by the straight-line segments $v_1 v_2$, $v_2 v_3$, etc., shown in he figure. We write the following set of four equations:

$$x = \sum_{i=1}^{6} n_i x_i \tag{9.13a}$$

$$y = \sum_{i=1}^{6} n_i y_i \tag{9.13b}$$

$$1 = \sum_{i=1}^{6} n_i \tag{9.13c}$$

$$n_i \geq 0 \tag{9.13d}$$

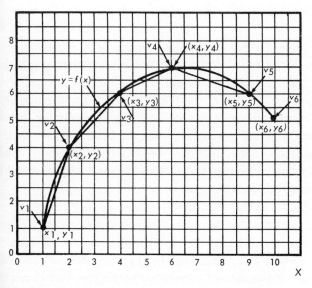

gure 9-2. *Piece-wise linear approximation to $y = f(x)$.*

In the above equations n_i's are parameters. The first two equation ((9.13a) and (9.13b)) give the values of x and y as a function of param eters n_i. Equation (9.13c) together with the requirement that *no mor than two of the* n$_i$*'s can be nonzero and these must be consecutive* wi restrict the x and y values to the line segments. The meaning of this wi become clear when we express Eqs. (9.13) numerically for $y = f(x)$ c Fig. 9-2. Reading the values of x_i and y_i from the figure, we have:

$$x = n_1(1) + n_2(2) + n_3(4) + n_4(6) + n_5(9) + n_6(10) \qquad (9.14a$$

$$y = n_1(1) + n_2(4) + n_3(6) + n_4(7) + n_5(6) + n_6(5) \qquad (9.14b$$

$$1 = n_1 + n_2 + n_3 + n_4 + n_5 + n_6 \qquad (9.14c$$

$$n_1, n_2, n_3, n_4, n_5, n_6 \geq 0 \qquad (9.14d$$

For example, a solution satisfying Eqs. (9.14c) and (9.14d) and th *consecutive* n$_i$ requirement will be $n_3 = 0.5$, and $n_4 = 0.5$ wit $n_1 = n_2 = n_5 = n_6 = 0$. This gives

$$x = 0.5(4) + 0.5(6) = 5$$
$$y = 0.5(6) + 0.5(7) = 6.5$$
$$(9.15$$

This is, as expected, a point on the line segment $v_3 v_4$. Another solu tion, for example, satisfying Eqs. (9.14c), (9.14d), and the *consecutiv* n$_i$ requirement will be $n_3 = 1.0$ with the rest of n_i's equal to zero. Thi corresponds to the point (4, 6) of Fig. 9-2, which is the end point of line segment. We emphasize again that in the linear programming formu lation n_i's represent a set of parameters that are treated as unknown Note that the condition specified by Eq. (9.14d) need not be incorporate in the formulation, since the method of linear programming deals onl with positive or zero values of the unknowns.

The above method can be extended to three or more dimension Three-dimensional relationships, geometrically expressed by surfaces, ca be replaced by a number of triangular planes approximating the surfac For example, the surface $f(x, y, z) = 0$ can be approximated by a numbe of triangular planes, and equations corresponding to Eqs. (9.13) can b written:

$$x = \sum_{i=1}^{n} n_i x_i \qquad (9.16\text{a})$$

$$y = \sum_{i=1}^{n} n_i y_i \qquad (9.16\text{b})$$

$$z = \sum_{i=1}^{n} n_i z_i \qquad (9.16\text{c})$$

$$1 = \sum_{i=1}^{n} n_i \qquad (9.16\text{d})$$

$$n_i \geq 0 \qquad (9.16\text{e})$$

In the above equations, x_i, y_i, and z_i represent the coordinates of the vertices of triangular planes, and n_i's are parameters. Together with the condition that no more than *three* (as compared with two of the previous case) of the n_i's can be nonzero and these must be associated with a triangular plane approximating the surface, Eq. (9.16d) will restrict the , y, and z values given by Eqs. (9.16a), (9.16b), and (9.16c) to the triangular plane. This becomes clear by taking the example of a surface approximated by only one triangular plane with vertices (1, 1, 1), 2, 2, 3), and (1, 2, 2), writing the equation of the plane passing through these, and showing that the points obtained from Eqs. (9.16) are located n this plane.

The general equation of a plane is written

$$Ax + By + Cz = 1 \qquad (9.17)$$

ubstituting the x, y, and z values at the vertices, we obtain a set of three quations and three unknowns, as follows:

$$\begin{aligned}
A + B + C &= 1 \\
2A + 2B + 3C &= 1 \\
A + 2B + 2C &= 1
\end{aligned} \qquad (9.18)$$

Solving the above equations for A, B, and C $(A = 1, B = 1, C = -1$ and substituting in Eq. (9.17), we obtain the equation of the plane passing through the vertices

$$x + y - z = 1 \qquad (9.19)$$

On the other hand, Eqs. (9.16) can be written for the triangular plane with the specified vertices as follows:

$$x = n_1(1) + n_2(2) + n_3(1)$$

$$y = n_1(1) + n_2(2) + n_3(2)$$

$$z = n_1(1) + n_2(3) + n_3(2) \qquad (9.20)$$

$$1 = n_1 + n_2 + n_3$$

$$n_1, n_2, n_3 \geq 0$$

For example, the values of $n_1 = 0.4$, $n_2 = 0.3$, and $n_3 = 0.3$ will satisfy the last two equations as well as being the n_i's associated with the triangular surface. The corresponding values of x, y, and z obtained from Eqs. (9.20) should be on the plane described by Eq. (9.19). Substituting the specified n_i's in Eqs. (9.20), we have

$$x = 1.3$$

$$y = 1.6$$

$$z = 1.9$$

which, indeed, satisfy Eq. (9.19) and hence are on the triangular plane.

The above principles have been used extensively in the formulation of the following problem.

FORMULATION OF A NUMERICAL PROBLEM

The formulation of separable programming problems is illustrated by two numerical examples. The problems selected are to find the optimum operating conditions in a power plant with three generators each with different power-fuel consumption characteristics [7].

Case 1

The first case is that of a power plant with three boiler-turbine-generator combinations with nonlinear generated power vs. fuel consumption characteristics. Two of the boilers use fuel oil and the third uses coal. The power plant load, combined outputs of the three generators, and the fuel costs are specified. The objective is to find the operating level of each generator to minimize total fuel costs.

The generated power-fuel consumption characteristics are given in Fig. 9-3. The λ_i's, μ_i's, and ν_i's specify points on the curves of the boiler-turbine-generator combinations λ, μ, and ν, respectively. The nonlinear curves are replaced by piecewise linear functions shown in the figure. For example, the curve of the boiler-turbine-generator combination ν is replaced by linear segments $\nu_1\nu_2$, $\nu_2\nu_3$, $\nu_3\nu_4$, and $\nu_4\nu_5$. Using these segments, the following equations for the generated power, fuel consumption, and the sum of n_i's can be written:

$$25n_1 + 23n_2 + 19.5n_3 + 16.7n_4 + 13n_5 = G_\nu \qquad (9.21a)$$

$$8.40n_1 + 7.86n_2 + 6.99n_3 + 6.45n_4 + 5.87n_5 = F_\nu \qquad (9.21b)$$

$$n_1 + n_2 + n_3 + n_4 + n_5 = 1.0 \qquad (9.21c)$$

where G_ν and F_ν are the generated power and coal consumption of the boiler-turbine-generator combination ν. Note that the last equation, (9.21c), along with the additional requirement that only two *consecutive* n_i's are to be nonzero will guarantee a solution on one of the specified line segments, as discussed previously. In the cases examined, the consecutive requirement was automatically satisfied, mainly because the generated power-fuel consumption curves were concave upward,* as noted by the following two observations. First, consider generator ν of Fig. 9-3 operating at 18 MW; the corresponding fuel consumption on line $\nu_3\nu_4$ is lower than on any other line passing through two ν's, for example, $\nu_2\nu_5$. Second, consider the main objective of the formulation

---

*This argument does not apply where the generated power-fuel consumption curves are concave downward—a condition rarely encountered in power plant operations. In these cases the additional provisions (specified in [4]) dealing with consecutive nonzero n_i and m_i values should be incorporated in the linear programming code.

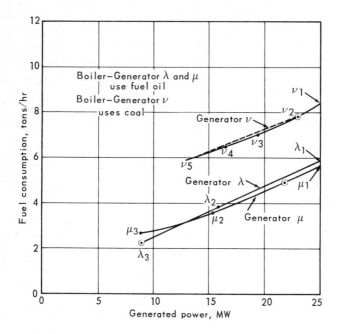

figure 9-3. *Generated power vs fuel consumption for Case 1.*

which is to minimize the amount of fuel (within the specified constraints). From these it is easy to see that at any operating level, the solution will seek the lowest obtainable fuel consumption, thereby satisfying the *consecutive* n_i requirement.

Similar equations can be written for the other two boiler-turbine-generator combinations λ and μ. The parameters associated with these are l and m, respectively. Note also that these use fuel oil as the boiler fuel.

After completing these calculations, the linear programming matrix of the problem can be constructed as given in Fig. 9-4. The matrix contains a listing of the unknowns, the constraint equations, and the functional, the latter representing the objective function to be minimized. The right-hand side (RHS) of the equations is given on the extreme right of the matrix. The linear programming matrix of Fig. 9-4 was solved by a computer code using the simplex method of solution.

In the matrix, G's and F's refer to generated electrical power and fuel consumption, respectively. The l's, m's, and n's specify parameters associated with the linearized curves of Fig. 9-3. Constraints 1 to 3 are

Linear programming matrix of Case 1.

	Generated power			Fuel consumption			Generator λ			Generator μ			Generator ν					RHS
Row no.	G_λ	G_μ	G_ν	F_λ	F_μ	F_ν	l_1	l_2	l_3	m_1	m_2	m_3	n_1	n_2	n_3	n_4	n_5	
1	-1						25	16	9									= 0.0
2				-1			5.90	3.84	2.25									= 0.0
3							1	1	1									= 1.0
4		-1								25	15.5	9						= 0.0
5					-1					5.59	3.60	2.64						= 0.0
6										1	1	1						= 1.0
7			-1										25	23	19.5	16.7	13	= 0.0
8						-1							8.40	7.86	6.99	6.45	5.87	= 0.0
9													1	1	1	1	1	= 1.0
10	1	1	1															= 54.0
Functional				12.6	12.6	10												= min
Answers	9.0	21.9	23.0	2.25	4.96	7.86			1.0		0.68	0.32	1.0					

Functional = 169.5

(Left margin labels: Unknowns; Constraints — rows 1–10; Functional)

figure 9-4. *Linear programming matrix of Case 1.*

353

for boiler-turbine-generator combination λ, 4 to 6 for μ, and 6 to 9 for ν. The constraint equation of row 10,

$$G_\lambda + G_\mu + G_\nu = 54.0 \text{ MW} \tag{9.22}$$

specifies the total load on the power plant. The functional represents the cost of fuel consumed by the plant. It is assumed that fuel oil costs 12.6 \$/ton and that coal cost 10 \$/ton. Thus, the functional becomes

$$12.6F_\lambda + 12.6F_\mu + 10F_\nu = \text{minimum} \tag{9.23}$$

as given on the last row of the matrix.

The solution of the linear programming matrix gave the numerical answers shown on Fig. 9-4. The corresponding operating levels are specified by (◯) on Fig. 9-3. Note that generators λ, μ, and ν operate at 9, 21.9, and 23 MW, respectively. The total fuel cost (functional) is 169.5 \$/hr.

Case 2

The second case is that of a power plant with two boiler-turbine-generator combinations, each capable of using a mixture of two types of fuel oil. The fuel consumption-load characteristics are nonlinear, while the fuel consumption-type of fuel characteristics are a linear function of the fuel blending ratios. The total electrical output of the power plant and the amount and type of fuel available at zero cost are specified. The object is to minimize the amount of purchased fuel.

The generated power-fuel consumption characteristics of this case are given in Fig. 9-5. Note that they can be considered as surface with fuel constituting the third dimension. The boiler-turbine-generator combinations λ and μ represent systems the boilers of which can use type-A or type-B fuel oil or any combination of these. For example, λ operating at 20 MW will use 5.5 tons per hour of type-B fuel oil or 5.65 tons per hour of a 1:1 ratio of type-A and type-B fuel oil. Connecting three consecutive λ's or μ's, we will obtain triangular planes for each of the two sets of surfaces. Thus, in the present case, we can have three nonzero values of l_i or m_i as opposed to two nonzero values in the previous cases. We should be careful, however, that the nonzero l_i and m_i values in the final solution represent a plane which is the closest approximation to the

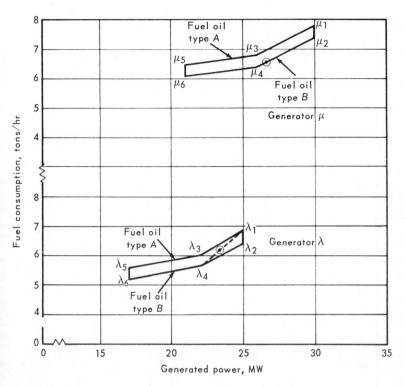

figure 9-5. *Generated power vs fuel consumption for Case 2.*

generated power-fuel consumption surface passing through the same l_i's or m_i's. From the results of the cases examined, because of the concave upward nature of the surfaces this condition was automatically satisfied.

After obtaining the coefficients from Fig. 9-5 we can set up the linear programming matrix of Fig. 9-6. The last two constraints specify the amount of available fuel at zero cost (10 tons of type-*B* fuel) and the desired power plant output (50.0 MW). The objective is to minimize the amount of purchased type-*A* fuel oil. $F_{A\lambda}$ and $F_{A\mu}$ represent type-*A* fuel used in generator-turbine-boiler combinations λ and μ, respectively. Similarly, $F_{B\lambda}$ and $F_{B\mu}$ represent type-*B* fuel used in generator-turbine-boiler combinations λ and μ.

The solution gave the results reported in Fig. 9-6 and the corresponding operating points specified ($\odot$) in Fig. 9-5. Note that generators λ and μ operate at approximately 23.5 and 26.5 MW, both below their rated capacities. The total type-*A* fuel oil consumption was 3.17 tons/hr.

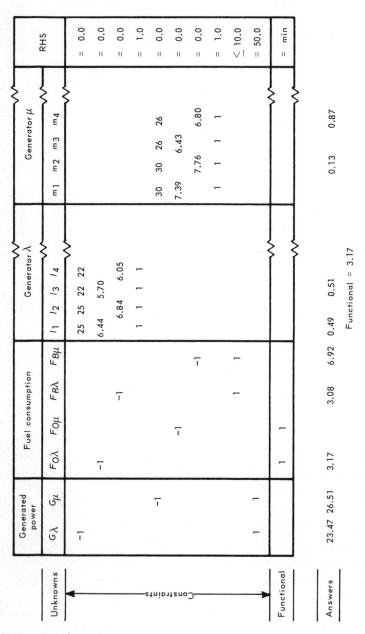

figure 9-6. *Linear programming matrix of Case 2.*

PROBLEMS

1. Obtain the solution of the following linear programming problem by the simplex method:

$$2x_1 - x_2 \leq 8$$

$$-x_1 + 3x_2 \leq 6$$

$$x_1 + 4x_2 = \text{maximum}$$

Show that the solution occurs at the intersection of the lines

$$2x_1 - x_2 = 8$$
$$-x_1 + 3x_2 = 6$$

$$\text{Answer:} \quad x_1 = 6, \ x_2 = 4.$$

2. Obtain the solution of the linear programming problem

$$10x_1 - 3x_2 \leq 30$$

$$-x_1 + 2x_2 \leq 4$$

$$x_1 + x_2 \geq 4$$

$$\text{Answer:} \quad x_1 = 4.235, \ x_2 = 4.117.$$

$$x_1 + x_2 = \text{maximum}$$

3. Write the equation of a plane passing through (1, 1, 1), (3, 4, 6), and (2, 3, 1) and equations corresponding to Eq. (9.20) of the text. Show that the point $n_1 = 0.2$, $n_2 = 0.2$, and $n_3 = 0.6$ is on the plane.

4. Formulate the separable programming matrix for the case of a power plant with two generator-turbine-boiler combinations λ and μ with the following characteristics

	Fuel consumption, tons/hour	Generated power, megawatts
	6	10
Generator λ	7	15
	10	20

	Fuel consumption, tons/hour	Generated power, megawatts
	5	10
Generator μ	6	15
	10	20

It is required to generate 35 megawatts, minimizing fuel consumption.

5. Obtain the solution of the linear programming problem

$$x_1 - x_2 + 2x_3 - x_4 = 2$$

$$2x_1 + x_2 - 3x_3 + x_4 = 6$$

$$x_1 + x_2 + x_3 + x_4 = 7$$

$$-2x_1 - x_2 + x_3 + x_4 = \text{maximum}$$

Answers: $x_1 = 3$, $x_2 = 0$
$x_3 = 1$, $x_4 = 3$

6. Obtain the solution of linear programming problem

$$2x_1 + x_2 \geq 2$$

$$x_1 + 3x_2 \leq 3$$

$$2x_2 \leq 4$$

$$3x_1 - x_2 = \text{maximum}$$

Answers: $x_1 = 3$
$x_2 = 0$

7. Write the logic for a digital computer program for the simplex method.

REFERENCES

1. Garvin, W. W.: "Introduction to Linear Programming," McGraw-Hill Book Company, New York, 1960.
2. Hovanessian, S. A., and E. L. Dougherty: Use of Linear Programming in Optimizing Oil Field Development, *Standard Oil Company of California Rept.* 725, July 18, 1961.

3. Wolfe, Philip: The Simplex Method for Quadratic Programming, *Econometrics,* vol. 27, p. 382, July 3, 1959.
4. Miller, C. E.: "The Simplex Method for Local Separable Programming," *Proc. Symp. Math. Programming, University of Chicago, June 18-22, 1962.*
5. Gass, Saul I.: "Linear Programming," McGraw-Hill Book Company, New York, 1964.
6. Hadley, G.: "Linear Programming," Addison-Wesley Publishing Company, Inc., Reading, Mass., 1962.
7. Hovanessian, S. A., and T. M. Stout: Optimum Fuel Allocation in Power Plants, *Trans. IEEE Power Apparatus Syst.,* June, 1963.

10 DYNAMIC PROGRAMMING

Dynamic programming has been successfully applied to optimizing returns from a set of engineering units, operating in series or in parallel, to produce certain outputs. In chemical engineering applications, the chemical process of production can be divided into a series of stages (units) where the output of one stage constitutes the input of the next stage. In industrial engineering applications, the production process can be considered to be accomplished in a series of units each requiring a certain amount of effort for a specified return. The problem of finding the most economical operating levels of electrical generators (each with different efficiency-load characteristics) in a power plant can be formulated by dynamic programming.

This discussion is intended to give the reader an understanding of the basic principles of dynamic programming. Numerical examples of parallel and series operation of production units are given. The results obtained for the parallel case are compared with the solutions obtained by the method of separable programming described in the previous chapter.

GENERAL DYNAMIC PROGRAMMING ANALYSIS

Consider a certain engineering production operation consisting of n units, as shown in Fig. 10-1. For convenience in calculation, number the

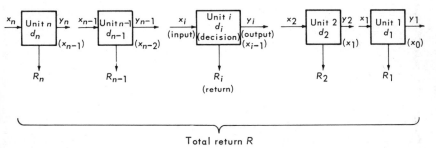

Total return R

figure 10-1. *Dynamic programming configuration of production units.*

last unit (on the right) as unit 1 and the first unit on the left as unit n, and assume the following notation:

x_i = input to unit i
y_i = output of unit i
d_i = decision required at unit i
R_i = return of unit i

In the above notation, x_i, y_i, and d_i can represent vectors. This means that input and output may consist of several components and there can be several types of decisions possible at each unit. Using the principle of dynamic programming, the total return R can be optimized (maximized or minimized) as follows.

Starting with unit 1, optimize the return of this unit R_1 with respect to its input x_1, output y_1, and decision d_1. Select the optimum operating points of unit 1 and combine this with operating points of unit 2. Optimize the combined returns of units 1 and 2. Select the corresponding optimum operating points. Combine these points with operating points of unit 3 and calculate the corresponding returns. Select the optimum operating points of unit 3 and the corresponding returns. Continue this process. This procedure is governed by the "principle of optimality" of dynamic programming [1]:

> No matter what the input (x_i) and decision (d_i) and resulting output (y_i) at unit i may be, the succeeding decision, d_{i+1}, must be made in such a way as to yield the optimum return from the $i+1$ units.

Let us formulate the mathematical expression of the above principle for the serial multistage process of Fig. 10-1. Let us further assume that the input to each stage is equal to the output of the previous stage, i.e.,

$$x_i = y_{i+1} \qquad (10.1)$$

For general stage i of the n-stage system of Fig. 10-1, we can write the expressions for the output and the return as a function of the input to stage i and the decision made at stage i.

$$x_{i-1} = t_i(x_i, d_i)$$
$$r_i = r_i(x_i, d_i) \qquad (10.2)$$

where x_{i-1} is the output of stage i (the input of stage $i-1$) and t_i and r_i represent functions of variables in parentheses. Since the output of every stage could be expressed as a function of its input and the stage decision, starting from stage n and proceeding to $n-1$, $n-2$, . . . and finally to $i+1$ which supplies the input of stage i, we can write

$$x_i = t_{i+1}(x_n, d_n, \ldots, d_{i+1}) \qquad (10.3)$$

which represents the input of stage i as a function of system input x_n and the decisions made up to stage i. Using this expression in the return (Eq. (10.2)), we can express

$$r_i = r_i(x_n, d_n, \ldots, d_i) \qquad (10.4)$$

In comparing Eqs. (10.3) and (10.4), we see that the decision of stage i, d_i, affects only the return of this stage and, of course, later stages.

The total n-stage system return R will be a function of individual returns from each stage; using Eq. (10.4), the total return can be expressed

$$R = g[r_n(x_n, d_n), r_{n-1}(x_n, d_n, d_{n-1}), \ldots, r_1(x_n, d_n, \ldots, d_1)] \tag{10.5}$$

The optimization problem will require the maximization or the minimization of the above return as a function of system input x_n and stage decisions $d_n, d_{n-1}, \ldots, d_1$. Using Eq. (10.2) for stage returns, the total return can be expressed also as

$$R = g[r_n(x_n, d_n), r_{n-1}(x_{n-1}, d_{n-1}), \ldots, r_i(x_1, d_1)] \tag{10.6}$$

Consider maximization of return R where

$$R = r_n(x_n, d_n) + r_{n-1}(x_{n-1}, d_{n-1}) + \cdots + r_1(x_1, d_1) \tag{10.7}$$

represents the sum of individual stage returns. Using the optimality principle and denoting by R_i the optimum i-stage return (stages 1 through i), we write

$$R_i = \max_{d_i}[r_i(x_i, d_i) + R_{i-1}] \tag{10.8}$$

The above is the *recursion* equation of dynamic programming. Starting from $i = 1$ and proceeding to $i = 2, 3, \ldots n$ the total return of i stages are maximized as a function of stage i decision d_i. Thus, the total n-stage system maximization problem is reduced to maximizing returns from 1, 2, 3, . . . , stages, respectively, and combining the cumulative return of $(i - 1)$ stages with the ith stage, and so on. As will be seen in the following examples, use of the methods of dynamic programming greatly reduces the number of calculations required in a particular optimization over the number which would be required if one had to choose the optimum from all of possible combinations of decisions.

NUMERICAL EXAMPLES

The following numerical examples will help to clarify the above principle.

Example of Parallel Operation

Consider a power plant with three boiler-turbine-generator combina
tions *A*, *B*, and *C*, with the cost vs. load curves given in Fig. 10-2. From
this figure the values of cost for various values of generated power, for

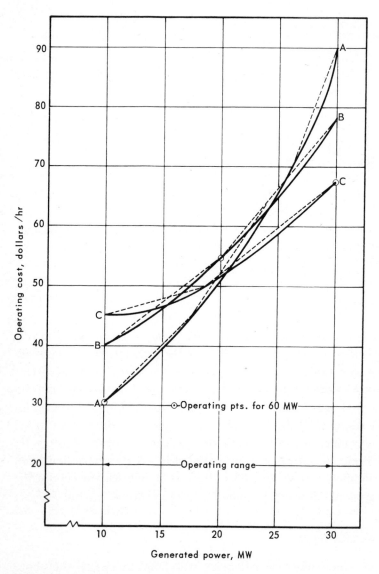

figure 10-2. *Operating cost vs. load for generators A, B, and C.*

ach of the three generators, can be obtained and tabulated. This abulation for increments of 5 MW is shown in Table 10-1. Suppose that ve want to generate a total of 60 MW in the most economical way and hat we proceed to formulate the problem using dynamic programming.

Table 10-1. Generated Power and Cost for Generators A, B, and C

Generator A		Generator B		Generator C	
Generated power, MW	Cost, $/hr	Generated power, MW	Cost, $/hr	Generated power, MW	Cost, $/hr
10	30	10	40	10	45
15	40	15	46	15	47
20	51	20	55	20	51
25	66	25	65	25	59
30	90	30	78	30	67

Let us take generator A operating at 10 MW and B operating at 10, 15, 20, 25, and 30 MW, respectively, and tabulate the values given in et 1 of Table 10-2. The first row of this set gives the cost of generating 20 MW, 10 by A and 10 by B, as 70 $/hr, which is the sum of cost igures from Table 10-1 (30 + 40 = 70 $/hr). In a similar manner, A operating at 10 MW and B at 15 MW will cost, from Table 10-1, 30 + 46 = 76 $/hr, as given in the second row of the first set in Table 10-2. Hence, we can complete the rest of the values of Table 10-2 with the aid of Table 10-1. Table 10-2 has, in fact, all of the possible operating combinations of generators A and B.

Up to this point we have merely tabulated values. At this time we utilize the governing principle of dynamic programming to reduce the number of computations when adding the output and costs of generator C to the respective values of generators A plus B as given in Table 10-2. An examination of Table 10-2 shows that there are several combinations of generators A and B that will result in the same amount of total power. For example, we can generate 35 MW by operating A at 10 and B at 25, as given in the fourth row of set 1; or A at 15 and B at 20, as given in the third row of set 2; or A at 20 and B at 15, as given in the second row of set 3; or A at 25 and B at 10, as given in the first row of set 4. But from all of the mentioned combinations for generating 35 MW, the combination of A at 10 and B at 25 with a total cost of 95 $/hr, as

Table 10-2. Operating Combinations of Generators A and B

Set	A operating at MW	B operating at MW	Cost $(A + B)$ $/hr	Total power $(A + B)$, MW
1	10	10	70	20*
	10	15	76	25*
	10	20	85	30*
	10	25	95	35*
	10	30	108	40
2	15	10	80	25
	15	15	86	30
	15	20	95	35
	15	25	105	40*
	15	30	118	45
3	20	10	91	30
	20	15	97	35
	20	20	106	40
	20	25	116	45*
	20	30	129	50*
4	25	10	106	35
	25	15	112	40
	25	20	121	45
	25	25	131	50
	25	30	144	55*
5	30	10	130	40
	30	15	136	45
	30	20	145	50
	30	25	155	55
	30	30	168	60*

given in the fourth row of set 1, is the most economical.* In a similar manner we can pick up the most economical ways of generating 20, 25, . . . , 60 MW from Table 10-2 as marked by asterisks. From now on we will be concerned only with the most economical combinations and costs of operating generators A and B and we can, in effect, drop all of the other values. Thus, in incorporating the generated power of generator C we use the combinations noted by asterisks in Table 10-2. These values are listed in Table 10-3. In this table we have two additional columns—one

---

*Note that 35 MW can also be generated at 95 $/hr by operating A at 15 and B at 20 as given in the third row of set 2. In case of ties either combination can be used for further study.

Table 10-3. Operating Combinations of Generators $A + B$ and C

Generated power $(A + B)$, MW	Cost $(A + B)$ $/hr	C, MW $60 - (A + B)$	Total cost
20	70		
25	76		
30	85	30	152*
35	95	25	154
40	105	20	156
45	116	15	163
50	129	10	174
55	144		
60	168		

for the load on generator C and the other for total cost. Since the total generated power as given in the problem is to be 60 MW, we calculate the load of generator C on this basis. For example, when $A + B$ are generating 20 MW, C should be generating $60 - 20 = 40$ MW. But this, from Fig. 10-1, is outside the operating range of C, so that no numbers are entered in the first row of the third and fourth columns. When $A + B$ are generating 40 MW, C will be generating $60 - 40 = 20$ MW; the cost will be $105 + 51 = 156$ $/hr, where 51 $/hr is the cost of C operating at 20 MW.

The rest of the values of Table 10-3 can be computed in the same manner. From the total cost column of this table it is seen that the 152 $/hr gives the most economical way of operating the generators for generating 60 MW of power. We see that the most economical operation corresponds to C operating at 30 MW and $A + B$ at 30 MW. Returning to Table 10-2, we obtain the operating levels of generators A and B which give the total of 30 MW. This value is given on the third row of set 1, and corresponds to A operating at 10 MW and B operating at 20 MW. These points (A at 10, B at 20, and C at 30) are marked on Fig. 10-1.

Now let us ask ourselves what we have gained by using dynamic programming. First, if we were to do this problem by considering all possible operating combinations of generators A, B, and C and corresponding costs, we would have $5 \times 5 \times 5 = 125$ combinations to consider (each generator can operate at 5 levels, that is, 10, 15, 20, 25, and 30). With dynamic programming we have reduced the number of calculations to 25 (in Table 10-2) + 5 (in Table 10-3) = 30. This is a reduction of over 75 percent. Second, we have followed a logical pattern for performing computations that can easily be programmed for a digital computer. The saving in computation will become more evident if we consider a finer grid for operating ranges of generators. For example, if we consider an increment of 2 instead of 5 MW we will have 11 load values for each generator, that is, 10, 12, 14, 16, 18, 20, 22, 24, 26, 28, and 30 MW. The total number of possible operating combinations will be $11 \times 11 \times 11 =$ 1,331. By dynamic programming the number of combinations will be reduced to $11 \times 11 + 11 = 132$ − a reduction of over 90 percent.

The solution given by Table 10-3 not only gives the most economical operating levels, but also gives other operating levels with increased cost. Sometimes it is desirable to operate at a level that is more practical than economical. For example, if we do not want to operate any of the generators at their rated capacity (30 MW), we might consider the solution corresponding to the total cost of 154 $/hr as given in Table 10-3. This solution results in operating A at 10, B at 25, and C at 25 MW; but of course this will cost us $154 − 152 = 2$ $/hr more than the optimum solution.

Now we ask ourselves whether if we had considered generators C and B first and then added generator A we would still get the same answers. The answer is that we will get the same optimum answer (A at 10, B at 20, C at 30 with a cost of 152 $/hr), but the other combinations will not be the same. This can be verified by taking generators C and B and completing a table similar to Table 10-2 and then incorporating generator A and completing a table similar to Table 10-3. This was done, and the final table of computations is shown in Table 10-4. A comparison of Tables 10-3 and 10-4 shows that optimum operating combinations

Table 10-4. Operating Combinations of Generators $C + B$ and A

Generated power $(C + B)$, MW	Cost $(C + B)$ $/hr	A, MW $60 − (C + B)$	Total cost
20 (C at 10, B at 10)	85		
25 (C at 15, B at 10)	87		
30 (C at 20, B at 10)	91	30	181
35 (C at 20, B at 15)	97	25	163
40 (C at 25, B at 15)	105	20	156
45 (C at 30, B at 15)	113	15	153
50 (C at 30, B at 20)	122	10	152*
55 (C at 30, B at 25)	132		
60 (C at 30, B at 30)	145		

and corresponding costs are the same in the two tables, but the inter-mediate operating combinations and costs are not the same. For example, from Table 10-4, the next lowest cost to the optimum is 153 $/hr while in Table 10-3 the corresponding value is 154 $/hr.

We solved the load allocation problem and presented the results in Table 10-3 for the total power generation of 60 MW. Using the first two columns of Table 10-3 we can find the most economical generating levels of the three generators for any other value of total generated power. The results of these calculations are illustrated in Fig. 10-3. From this figure, for example, we observe that the most economical ways of generating 75 MW and 80 MW are operating A at 20, B at 25, and C at 30 at a cost of 182.5 $/hr and operating A at 20, B at 30, and C at 30 at a cost of 195 $/hr, respectively. This figure can also be used for obtaining operating levels that are not multiples of 5. For example, if we want to generate 78 MW at minimum cost, this value can be obtained from Fig. 10-3 by operating A at 20, B at 30, and increasing the operating level of C to 25 + 3 = 28 MW. The cost, by linear extrapolation, will

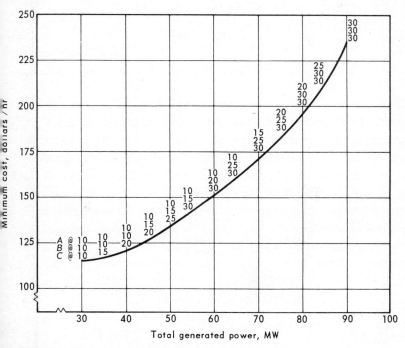

figure 10-3. *Most economical power generation levels of generators A, B, and C for required total power generation.*

be approximately $182.5 + 3/5 (192 - 182.5) = 190$ \$/hr. Of course, a more accurate answer can be obtained by solving the problem with load increments of 1 MW.

To compare the above solutions with the solutions obtained by other methods, we will solve the problem by the method of separable programming.

Separable Programming Formulation

The problem described above can also be solved by separable programming, as discussed in the previous chapter and in [2]. In formulating the problem for this method we need to substitute piecewise linear approximations for the nonlinear cost-load curves of Fig. 10-2. This approximation is shown by dashed lines in the figure. The equations for generated power and cost can be written in terms of "special variables" which incorporate the piecewise linear approximations of the nonlinear curves. The equations for generator A can be written as follows:

$$-P_A + 10\lambda_1 + 17.5\lambda_2 + 26.25\lambda_3 + 30\lambda_4 = 0$$

$$-C_A + 30\lambda_1 + 45\lambda_2 + 70\lambda_3 + 90\lambda_4 = 0 \qquad (10.9)$$

$$\lambda_1 + \lambda_2 + \lambda_3 + \lambda_4 = 1$$

where λ_1, λ_2, λ_3, and λ_4 are the special variables and their coefficients in the first two equations are the coordinates of the intersections of line segments describing the curve of generator A in Fig. 10-2. The first two equations give the amount and cost of generated power, respectively. The last equation, with the additional condition that only two of the λ's can be nonzero and that these must be consecutive, will guarantee a solution on one of the specified line segments. Equations similar to Eq. (10.9) can be written for generators B (with special variables μ) and C (with special variables γ). The separable programming matrix of the problem is shown in Fig. 10-4. Rows 1 through 9 of the matrix give constraint equations for generators A, B, and C, while row 10 specifies the total required amount of generated power, i.e.,

$$P_A + P_B + P_C = 60 \qquad (10.10)$$

In Fig. 10-4 the designation RHS specifies right-hand side of equations.

Row	CA	CB	CC	PA	PB	PC	λ1	λ2	λ3	λ4	μ1	μ2	μ3	γ1	γ2	γ3	RHS
1				-1			10	17.5	26.25	30							
2	-1						30	45	70	90							
3							1	1	1	1							1
4					-1						10	20	30				
5		-1									40	55	78				
6											1	1	1				1
7						-1								10	18.5	30	
8			-1											45	50	67	
9														1	1	1	1
10				1	1	1											60
Functional	1.0	1.0	1.0														min
Answers	30	55	67	10	20	30	1.0					1.0				1.0	

Functional = 152 dollars/hr

figure 10-4. *Separable programming matrix of the allocation problem.*

The solution of the problem by the simplex method of linear programming is shown at the bottom of the figure. This solution results:

Generated power, MW	Cost, $/hr
$P_A = 10$	30
$P_B = 20$	55
$P_C = 30$	67
Total 60	152

Note that the above solution is exactly the same as obtained by dynamic programming.

The method of dynamic programming as applied to the above problem is simple to follow and the calculations involved are logical and can be performed by digital computers with a small amount of programming effort.   The accuracy of the solution can be improved, if necessary, by taking smaller increments for the generated power. This will increase the number of calculations to be performed although the logical sequence of calculations will remains the same.

The solution obtained by separable programming will depend on the accuracy of representation of cost-load curves by piecewise linear segments. We note that this accuracy can be easily increased by taking more linear segments to represent a given curve.   This in turn will increase the number of special variables, but fortunately an increase in the number of variables does not appreciably increase the computation time. Hence, by separable programming we have a greater control on the accuracy of the solution.   In addition, solutions can be obtained for any value of required total generated power whereas this will require finer increments of generated power in dynamic programming. The separable programming method, of course, will require a computer code (usually the simplex method) for solution, while dynamic programming does not involve any special method.

Example of Series Operation

Consider the three-stage production process of Fig. 10-5. The products manufactured at stage 1, x_0, have a market value of $10 each. These products can be manufactured from input products x_1, using one of three processes. The input-output relationships of these processes are given in Fig. 10-6 as a function of types of process (decision) $d_1, d_2,$ and d_3.

figure 10-5. *Series production process.*

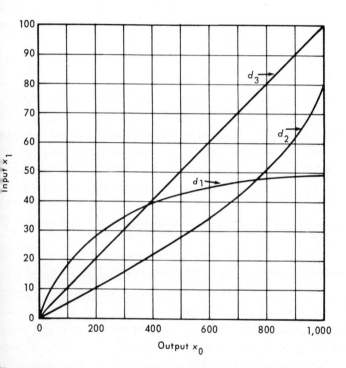

figure 10-6. *Input-output relationship of stage 1 with decision, d,*
is parameter.

These decisions have the following cost schedule:

Cost of processing each unit of x_1 with decision $d_1 = \$10$
Cost of processing each unit of x_1 with decision $d_2 = \$5$
Cost of processing each unit of x_1 with decision $d_3 = \$5$
Cost of implementing decision $d_1 = \$500$
Cost of implementing decision $d_2 = \$700$
Cost of implementing decision $d_3 = \$600$

The input-output relationship of stage 2 is given as a function of the
type of process (decision) in Fig. 10-7. In stage 2 two types of decisions,
with the following cost schedule, are possible:

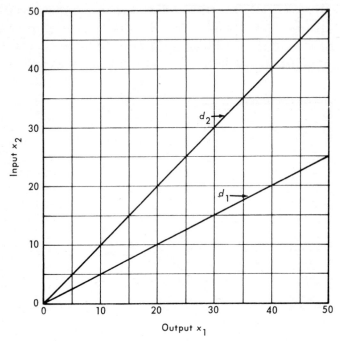

figure 10-7. *Input-output relationship of stage 2 with decision, d, as parameter.*

Cost of processing each unit of x_2 with decision d_1 = $10
Cost of processing each unit of x_2 with decision d_2 = $5
Cost of implementing decision d_1 = $600
Cost of implementing decision d_2 = $500

In stage 3 two types of decision are possible. The cost schedule of these decisions is as follows:

Cost of processing each unit of x_2 with decision d_1 = $10
Cost of processing each unit of x_2 with decision d_2 = $20
Cost of implementing decision d_1 = $500
Cost of implementing decision d_2 = $100

With the above data it is desired to obtain the optimum decisions of each stage for producing 100, 200, 300, . . . , 1,000 units of x_0 for maximum return.

For the solution of the problem consider stage 1 and obtain the values of input x_1 for each value of x_0 and decision d_1. These values are given in Table 10-5. As an example of calculating the values of this table, consider output x_0 of 500 units and decision d_2. From Fig. 10-6, an output

Table 10-5. Input and Return of Stage 1

Output x_0	Decision d_1	Decision d_2	Decision d_3
	Input x_1/return, $	Input x_1/return, $	Input x_1/return, $
100	19/ 310	5/ 275	10/ 350*
200	28/1,220	10/1,250	20/1,300*
300	35/2,150	16/2,220	30/2,250*
400	40/3,100	21/3,195	40/3,200*
500	43/4,070	27/4,165*	50/4,150
600	45/5,050	34/5,130*	60/5,100
700	47,6,030	41/6,095*	70/6,050
800	48/7,020	50/7,050*	80/7,000
900	49/8,010*	62/7,990	90/7,950
1,000	50/9,000*	80/8,900	100/8,900

of 500 units corresponds to an input x_1 of 27 units using decision d_2 process. Since the selling price of each unit of x_0 is $10, the total return of stage 1 will be

$$R = 500 \times 10 - (27 \times 5 + 700) = \$4,165$$

where 5 is the cost of processing each unit x_1 with decision d_2, and 700 is the initial cost of implementing decision d_2. The stage return and the number of inputs required are separated by a slash (/) in Table 10-5. After completing the values required in this table we choose, for each output x_0, the most profitable operating decision-highest return. These values are marked with an asterisk. From now on we need only concern ourselves with these values.

Moving to a two-stage process consisting of stages 1 and 2, we tabulate the values of x_0 and the values of x_1 corresponding to the most profitable operation of stage 1. Columns 1, 2, and 3 of Table 10-6 represent these values along with the corresponding return of stage 1. Columns 4 and 5 give the corresponding number of input units x_2 obtained for the output units x_1 of stage 2 obtained from Fig. 10-7. These columns also include the combined two-stage return. As an example of calculations, consider

Table 10-6. Input and Returns of Two-stage Process

Output x_0 units	Input to Stage 1 or output to stage 2, x_1	Stage 1 return, $	Decision d_1		Decision d_2	
			Stage 2 input x_2	Combined stage 1 & 2 return, $	Stage 2 input x_2	Combined stage 1 & 2 return, $
100	10	350	5	-300	10	-200*
200	20	1,300	10	600	20	700*
300	30	2,250	15	1,500	30	1,600*
400	40	3,200	20	2,400	40	2,500*
500	27	4,165	14	3,425	27	3,530*
600	34	5,130	17	4,360	34	4,460*
700	41	6,095	21	5,285	41	5,390*
800	50	7,050	25	6,200	50	6,300*
900	49	8,010	25	7,160	49	7,265*
1,000	50	9,000	25	8,150	50	8,250*

he calculation involved for x_0 of 800 and decision d_2 as represented in olumn 5 of Table 10-6. For an x_0 of 800, the corresponding x_1 from able 10-5 is 50 with a return of \$7,050. From Fig. 10-7, an $x_1 = 50$ orresponds to an input to stage 2, x_2, of 50 using decision d_2. The orresponding return of the two-stage process is calculated as follows:

$$R = 7,050 - (50 \times 5 + 500) = \$6,300$$

where 5 is the cost of processing each unit of x_2 and 500 is the cost of mplementing decision d_2. These costs are subtracted from the one-stage eturn of \$7,050. After completing Table 10-6 the returns are optimized with respect to decisions and distinguished by an asterisk.

The return of the three-stage process can be computed in a similar manner and a table of values obtained. As was the case with the values of Table 10-6, only the asterisked values of Table 10-6 will be used in he compilation of the three-stage process (Table 10-7). This table gives he most profitable two-stage values in columns 1 through 4 as obtained from Table 10-6. The three-stage returns are calculated for decisions d_1 nd d_2 of the third stage. For example, for a value of $x_0 = 700$ lecision d_1 will result in the three-stage return:

$$R = 5,390 - (41 \times 10 + 500) = \$4,480$$

where 41 is the number of stage-three outputs required, 10 is the cost of producing each unit, and 500 is the initial cost of implementing

Table 10-7. Three-stage Return

x_0 units	x_1 units	x_2 units	Two-stage return, \$	Decision d_1 Three-stage return, \$	Decision d_2 Three-stage return, \$
100	10	10	-200	-800	-500*
200	20	20	700	0	200*
300	30	30	1,600	800	900*
400	40	40	2,500	1,600	1,600*
500	27	27	3,530	2,760	2,890*
600	34	34	4,460	3,620	3,680*
700	41	41	5,390	4,480*	4,470
800	50	50	6,300	5,300*	5,200
900	49	49	7,265	6,275*	6,185
,000	50	50	8,250	7,250*	7,150

decision d_1. The most profitable three-stage returns are marked by asterisks. Thus Tables 10-5 through 10-7 give the most profitable decisions
for producing quantities of x_0. The results can be combined and written
in a single table, as in Table 10-8. This table gives the required decision
at each stage and the resulting profit for producing various quantities of
x_0. From this table it is seen that manufacturing 100 units of x_0 will
incur a loss of $500, while manufacturing 1,000 units will bring a profit
of $7,250.

PROBLEMS

1. Find the optimum level of operating generators A, B, and C of the
   example problem for generating 60 MW by taking combinations of
   generators A and C first, and combining the results with generator
   B.
2. Find the optimum level of operating generators A, B, and C of the
   example problem for generating 80 MW of total power. Check your
   answer with the value given in Fig. 10-3.
3. Repeat the above problem for 70 MW of total generated power.
4. In the three-stage production problem, assume the following cost
   figures for the first stage:

Table 10-8. The Overall Decision Process and Return of the Three-stage Series Example

Output x_0 units	Stage 1			Stage 2		Stage 3		Overall Profit, $
	d_1	d_2	d_3	d_1	d_2	d_1	d_2	
100			✓		✓		✓	−500
200			✓		✓		✓	200
300			✓		✓		✓	900
400			✓		✓		✓	1,600
500		✓			✓		✓	2,890
600		✓			✓		✓	3,680
700		✓			✓	✓		4,480
800		✓			✓	✓		5,300
900	✓				✓	✓		6,275
1,000	✓				✓	✓		7,250

Cost of processing each unit of x_1 with decision $d_1 = \$8$
Cost of processing each unit of x_1 with decision $d_2 = \$4$
Cost of processing each unit of x_1 with decision $d_3 = \$2$
Cost of implementing decision $d_1 = \$500$
Cost of implementing decision $d_2 = \$500$
Cost of implementing decision $d_3 = \$400$

and obtain the optimum production decisions as was done in the example problem.

5. Taking the cost figures and curves of the three-stage production problem, calculate the most economical decisions for producing x_0's in 50, 150, 250, 350, and 450 units.

6. In the three-stage production problem, drop decision d_3 from consideration in stage 1. With only decisions d_1 and d_2 remaining, calculate three-stage return for producing 100, 200, 300, and 400 units of x_0.

REFERENCES

1. Bellman, R. E., and S. E. Dreyfus: "Applied Dynamic Programming," chap. 1, Princeton University Press, Princeton, N. J., 1962.

2. Miller, C. E.: "The Simplex Method for Local Separable Programming," *Proc. Symp. Math. Programming, University of Chicago, June 18-22, 1962.*

3. Nemhauser, G. L.: "Introduction to Dynamic Programming," John Wiley & Sons, Inc., New York, 1966.

APPENDIX
DIFFERENTIAL EQUATIONS

DEFINITIONS

The term *ordinary differential equation* is given to an equation containing one or more *ordinary derivatives or differentials*; if the derivatives are *partial derivatives,* the equation is called a *partial differential equation.* For example,

$$\frac{d^2x}{dt^2} + A\frac{dx}{dt} + Bx = 0 \qquad\qquad (A.1)$$

is an ordinary differential equation, while the equation

$$\frac{\partial^2 V}{\partial x^2} + \frac{\partial^2 V}{\partial y^2} + \frac{\partial^2 V}{\partial z^2} = 0 \qquad\qquad (A.2)$$

is a partial differential equation. Note that in Eq. (A.1) the variable x is a function of t only, while in Eq. (A.2) the variable V is a function of x, y, and z.

The *order* of a differential equation is the order of the highest derivative involved in the equation. For example, Eqs. (A.1) and (A.2) are both of the *second* order.

The general solution of an *nth-order* differential equation contains n *arbitrary* constants. Consider the solution of the first-order differential equation

$$\frac{dx}{dt} = 1 + 2x \qquad (A.3)$$

Separation of variables results in

$$\frac{dx}{1 + 2x} = dt \qquad (A.4)$$

Integration gives

$$\frac{1}{2} \ln(1 + 2x) = t + C \qquad (A.5)$$

in which C is the arbitrary constant. The constant C can be evaluated using the initial condition. For $x = 1/2$ at $t = 0$, we get

$$C = \frac{1}{2} \ln 2 \qquad (A.6)$$

The solution (A.5) after this substitution results in

$$x = e^{2t} - \frac{1}{2} \qquad (A.7)$$

The general solution of the second-order differential equation

$$\frac{d^2x}{dt^2} + \omega^2 x = 0 \qquad (A.8)$$

is given by

$$x = A \sin\omega t + B \cos\omega t \tag{A.9}$$

where A and B are the two arbitrary constants. The fact that Eq. (A.9) is a solution of Eq. (A.8) can be verified by obtaining the second derivative of x with respect to time and substituting in the differential equation (A.8). Given the initial conditions, the constants A and B can be evaluated.

A *linear* differential equation is one in which the dependent variable and its derivatives appear *only* in the first power. The differential equations

$$\frac{d^2y}{dx^2} + 3\frac{dz}{dx} + 2y = 0 \tag{A.10}$$

and

$$\frac{d^2y}{dx^2} + (\sin x)\frac{dy}{dx} + (\cos x)\, y = 0 \tag{A.11}$$

are *linear,* while the differential equations

$$\frac{d^2y}{dx^2} + 3\left(\frac{dy}{dx}\right)^2 + 2y = 0 \tag{A.12}$$

and

$$\frac{d^2y}{dx^2} + 3\frac{dy}{dx} + 2y^3 = 0 \tag{A.13}$$

are *nonlinear.*

The *principle of superposition* of solutions applies in the case of linear differential equations. This principle can be stated as follows. If $y = f(x)$ and $y = g(x)$ are solutions of a linear differential equation, then

$y = f(x) + g(x)$ is also a solution. It can be seen by substitution that $y = Ae^{-x}$, where A is constant, is a solution of Eq. (A.10). Again by substitution, it can be seen that $y = Be^{-2x}$, where B is a constant, is also a solution of Eq. (A.10). By the superposition principle $y = Ae^{-x} + Be^{-2x}$ contains two constants A and B and will be the complete solution of the second-order linear differential equation (A.10).

LINEAR EQUATIONS OF THE FIRST ORDER

Consider the first-order differential equation

$$\frac{dy}{dx} + Py = Q \tag{A.14}$$

where P and Q are functions of independent variable x. We wish to obtain the solution y in terms of x. Equation (A.14) can be written as

$$dy + Pydx = Qdx \tag{A.15}$$

Multiplying Eq. (A.15) by R, an unknown function of x, we get

$$Rdz + yRPdx = RQdx \tag{A.16}$$

Comparing this equation with the identity

$$Rdy + ydR = d(Ry) \tag{A.17}$$

we get

$$dR = RPdx \tag{A.18}$$

and

$$d(Ry) = RQdx \tag{A.19}$$

From Eq. (A.18) we obtain

$$\frac{dR}{R} = Pdx \quad , \quad R = \exp\left(\int Pdx\right) \tag{A.20}$$

From Eq. (A.19) we obtain

$$Ry = \int RQdx + C \tag{A.21}$$

where C is an arbitrary constant. Since Eq. (A.20) gives the value of R as a function of x, Eq. (A.21) will give the solution y as a function of x and the arbitrary constant C.

As an example of the above, consider the solution of the differential equation

$$\frac{dy}{dx} - \frac{2}{x}y = 2x^3 \tag{A.22}$$

Comparing this to Eq. (A.14) we have

$$P = -\frac{2}{x} \quad, \quad Q = 2x^3 \tag{A.23}$$

which results in

$$R = \exp\left(\int -\frac{2}{x}dx\right) = \exp(\ln x^{-2}) = x^{-2} \tag{A.24}$$

Using this in Eq. (A.21) will give

$$x^{-2}y = \int x^{-2} \cdot 2x^3 dx + C \tag{A.25}$$

$$= x^2 + C$$

or

$$y = x^4 + Cx^2 \tag{A.26}$$

where C is the arbitrary constant.

LINEAR EQUATIONS WITH CONSTANT COEFFICIENTS

Consider the linear differential equation

$$a_0 D^n y + a_1 D^{n-1} y + \cdots + a_{n-1} Dy + a_n y = R \tag{A.27}$$

where a is a constant and D is the differential operator, $D = d/dx$, and R is a function of the independent variable x. Equation (A.27) can be written

$$f(D) y = R$$
$$f(D) = a_0 D^n + a_1 D^{n-1} + a_2 D^{n-2} + \cdots + a_n \tag{A.28}$$

SOLUTION FOR $R = 0$

Consider the case of Eq. (A.27) when $R = 0$ and $n = 1$. This will result in

$$(D - r_n) y = 0 \tag{A.29}$$

where $r_n = a_n/a_{n-1} = a_1/a_0$. Equation (A.29) can be written

$$\frac{dy}{dx} = r_n y \tag{A.30}$$

resulting in the solution

$$y = C_n \exp(r_n x) \tag{A.31}$$

Generalizing this, if the roots of the equation $f(D) = 0$ are distinct and equal to $r_1, r_2, \ldots, r_n$ then the solution y becomes

$$y = C_1 e^{r_1 x} + C_2 e^{r_2 x} + C_3 e^{r_3 x} + \cdots + C_n e^{r_n x} \tag{A.32}$$

As an example of this consider the second-order differential equation

$$2 \frac{d^2 y}{dx^2} + \frac{dy}{dx} - y = 0 \tag{A.33}$$

The roots of

$$f(D) = 2D^2 + D - 1$$
$$= (2D - 1)(D + 1) \qquad \text{(A.34)}$$

are $r_1 = 1/2, r_2 = -1$. This results in the solution

$$y = C_1 e^{1/2x} + C_2 e^{-x} \qquad \text{(A.35)}$$

If the roots of the equation $f(D) = 0$ are multiple roots with multiplicity n such that the differential equation becomes

$$(D - r)^n y = 0 \qquad \text{(A.36)}$$

then the solution can be shown to be

$$y = (C_1 + C_2 x + C_3 x^2 + \cdots + C_n x^{n-1}) e^{rx} \qquad \text{(A.37)}$$

As an example of this consider the differential equation

$$\frac{d^2 y}{dx^2} - 6 \frac{dz}{dx} + 9y = 0 \qquad \text{(A.38)}$$

resulting in an $f(D)$ of

$$f(D) = D^2 - 6D + 9$$
$$= (D - 3)^2 \qquad \text{(A.39)}$$

with multiple roots $r_1 = r_2 = 3$. The solution of Eq. (A.38) according to Eq. (A.37) will be

$$y = (C_1 + C_2 x) e^{3x} \qquad \text{(A.40)}$$

If the roots of the equation $f(D) = 0$ are complex conjugate roots $m_1 = a + jb$ and $m_2 = a - jb$, then the solution

$$y = C_1 e^{m_1 x} + C_2 e^{m_2 x} \tag{A.41}$$

can also be written as

$$y = K_1 e^{ax} \sin bx + K_2 e^{ax} \cos bx \tag{A.42}$$

where C_1, C_2, K_1 and K_2 are arbitrary constants.
   For example, the differential equation

$$\frac{d^2 y}{dx^2} + 6\frac{dy}{dx} + 25y = 0 \tag{A.43}$$

with the roots of $D^2 + 6D + 25 = 0$ being equal to $-3 + 4j$ and $-3 - 4j$ will have a solution $y = K_1 e^{-3x} \cos 4x + K_2 e^{-3x} \sin 4x$.

Solution for $R \neq 0$

For $R \neq 0$ the differential equation (A.27) can be written in the form

$$f(D) y = R \tag{A.44}$$

The solution of the differential equation (A.44) in this case will be the sum of two solutions y_1 and y_2 where y_1 is the solution of

$$f(D) y_1 = 0 \tag{A.45}$$

and y_2 is the solution of

$$f(D) y_2 = R \tag{A.46}$$

In Eq. (A.45) the value of R is set equal to zero. The solution of Eq. (A.45), y_1, is called the *complementary* function while y_2, the solution of Eq. (A.46), is called a *particular* solution. The solution of Eq. (A.45) can be obtained by methods described previously. The solution of Eq. (A.46) can be obtained by assuming y_2 to be proportional to the sum of functions appearing in R (for example, x^p, e^{px}, $\sin qx$, $\cos ux$) and their derivatives as will be described in the following examples.

Example 1. Solve the differential equation

$$(D^2 + 2D + 1)y = 4e^x \tag{A.47}$$

The equation corresponding to Eq. (A.45) can be written

$$(D^2 + 2D + 1)y_1 = 0 \tag{A.48}$$

This results in the complementary solution

$$y_1 = (C_1 + C_2 x)e^{-x} \tag{A.49}$$

The particular solution y_2 of Eq. (A.46) is obtained by assuming an exponential function of the type given in Eq. (A.47), $R = 4e^x$,

$$y_2 = Ae^x \tag{A.50}$$

Obtaining appropriate derivatives and substituting in Eq. (A.47), we obtain

$$Ae^x + 2Ae^x + Ae^x = 4e^x \tag{A.51}$$

which results in $A = 1$. Thus the complete solution of Eq. (A.47) becomes

$$y = y_1 + y_2 = (C_1 + C_2 x)e^{-x} + e^x \tag{A.52}$$

where C_1 and C_2 are the two arbitrary constants of the second-order differential equation (A.47).

Example 2. Solve the second-order differential equation

$$\frac{d^2 y}{dx^2} - \frac{dy}{dx} = 3x^2 - 4x + 5 + 2e^x + \sin x \tag{A.53}$$

The complementary solution is obtained from the equation

$$(D^2 - D)y_1 = 0 \tag{A.54}$$

as

$$y_1 = C_1 + C_2 e^x \tag{A.55}$$

Considering the right side of Eq. (A.53) the particular solution y_2 will contain functions of the same form,

$$x^2, x, \text{constant}, e^x, \sin x \tag{A.56}$$

We note that the constant term also appears in solution y_1 of Eq. (A.55) as C_1. For this reason the powers of x appearing in Eq. (A.56) should be multiplied by x,

$$y_2 = Ax^3 + Bx^2 + Cx + \cdots \tag{A.57}$$

where A, B, and C are constants. The term e^x of Eq. (A.56) also appears in the complementary solution y_1 of Eq. (A.55). This term should also be multipled by x resulting in

$$y_2 = Ax^3 + Bx^2 + Cx + Exe^x + \cdots \tag{A.58}$$

Adding the term $\sin x$ and its derivative $\cos x$, we can write the particular solution as

$$y_2 = Ax^3 + Bx^2 + Cx + Exe^x + F \sin x + G \cos x \tag{A.59}$$

Obtaining the derivatives of the above equation and substituting in Eq. (A.53), we will obtain an equation containing the constants $A, B, C, \ldots$ and the functions appearing in Eq. (A.59). Equating the coefficients of the like terms, we can solve the resulting simultaneous equations for the values of the constants,

$$A = -1, B = -1, C = -7, E = 2, F = -\tfrac{1}{2}, G = \tfrac{1}{2} \tag{A.60}$$

The complete solution of differential equation (A.53) now becomes

$$y = C_1 + C_2 e^x - x^3 - x^2 - 7x + 2xe^x - \tfrac{1}{2} \sin x + \tfrac{1}{2} \cos x \tag{A.61}$$

where C_1 and C_2 are constants.

SYSTEM OF LINEAR EQUATIONS

The sets of simultaneous differential equations can also be solved using the methods discussed above. Consider the set of differential equations

$$\frac{dy}{dx} + \frac{dz}{dx} = 4y + 1$$

$$\frac{dy}{dx} + z = 3y + x^2 \tag{A.62}$$

where y and z are functions of the independent variable x. Equation (A.62) can be written as

$$(D - 4)y + Dz = 1$$

$$(D - 3)y + z = x^2 \tag{A.63}$$

Multiplying the second equation by D and subtracting from the first we get

$$(D^2 - 4D + 4)y = (D - 2)^2 y = 2x - 1 \tag{A.64}$$

The complementary and particular solutions of Eq. (A.64) are as follows:

$$y_1 = (C_1 + C_2 x)e^{2x} \tag{A.65}$$

and

$$y_2 = \frac{x}{2} + \frac{1}{4} \tag{A.66}$$

This results in

$$y = (C_1 + C_2 x)e^{2x} + \frac{x}{2} + \frac{1}{4} \tag{A.67}$$

Solving the second of Eqs. (A.63) for z and substituting the value of y and its derivative from Eq. (A.67), we get

$$z = (C_1 - C_2 + C_2 x)e^{2x} + x^2 + \frac{3}{2}x + \frac{1}{4} \tag{A.68}$$

As a second example, consider the solution of the sets of equations

$$\frac{d^2x}{dt^2} + 2\left(\frac{dy}{dt} - a\right) = 0$$

$$\frac{d^2y}{dt^2} - 2\left(\frac{dx}{dt} - b\right) = 0$$

(A.69)

The above equations can be written

$$D^2x + 2Dy = 2a$$

$$2Dx - D^2y = 2b$$

(A.70)

Solving the above for x, we obtain

$$(D^3 + 4D)x = 4b$$

(A.71)

The solution of Eq. (A.71) is

$$x = C_1 + C_2 \sin 2t + C_3 \cos 2t + bt$$

(A.72)

Using the second derivative of Eq. (A.72) in the first of Eqs. (A.69) we get

$$Dy = 2C_2 \sin 2t + 2C_3 \cos 2t + a$$

(A.73)

Integrating this we obtain

$$y = -C_2 \cos 2t + C_3 \sin 2t + at + C_4$$

(A.74)

Equations (A.72) and (A.74) with the four constants C_1, C_2, C_3, and C_4 constitute the solution of the two second-order differential equations (A.69).

PROBLEMS

1. Show that $y = A \sin\omega x + B \cos\omega x$ is the solution of the differential equation

$$\frac{d^2y}{dx^2} + \omega^2 x = 0$$

2. Show that $y = A \sin(\omega x + B)$ and $y = C \cos(\omega x + D)$ are the solutions of the differential equation

$$\frac{d^2y}{dx^2} + \omega^2 x = 0$$

3. Solve the differential equation

$$x\frac{dy}{dx} + 3y = x^2 + 1$$

and check your solution by substituting in the differential equation.

4. Solve the differential equation

$$(D^4 - 6D^2 + 1)\,y = 0$$

5. Obtain the solution of differential equation

$$\frac{d^3y}{dx^3} + 4\frac{dy}{dx} = 0$$

Answer: $y = C_1 + C_2 \cos 2x + C_3 \sin 2x$

6. Obtain the solution of the differential equation

$$(D^2 + 7D + 12)\,y = 2e^{2x}$$

Answer: $y = C_1 e^{-3x} + C_2 e^{-4x} + \frac{1}{15}e^{2x}$

7. Solve the differential equation

$(D^2 + D - 12)y = xe^x$

Answer: $y = C_1 e^{3x} + C_2 e^{-4x} - \frac{1}{10} xe^x - \frac{3}{100} e^x$

8. Solve the differential equation

$(D^2 - 2D + 1)y = 2xe^x$

Hint: The particular solution contains the terms $x^3 e^x$ and $x^2 e^x$.

9. Solve the differential equation

$(D^4 + 4D^3)y = x^2$

Answer: $y = C_1 e^{-4x} + C_2 + C_3 x + C_4 x^2 + \frac{x^3}{192} - \frac{x^4}{192} + \frac{x^5}{240}$

10. Solve the differential equation

$(D^2 + 4)y = \sin 2x$

Answer: $y = C_1 \sin 2x + C_2 \cos 2x - \frac{1}{4} x \cos 2x$

11. Solve the set of simultaneous differential equations

$\frac{d^2 y}{dt^2} - 6x + 2y = 0$

$\frac{d^2 y}{dt^2} - 7y + 3x = 0$

12. Solve the simultaneous set of differential equations

$\frac{dy}{dx} - y - z = 2 \cos 2x$

$\frac{dz}{dx} - 3y + z = 0$

REFERENCES

1. Boyce, William E., and Richard C. DiPrima: "Elementary Differential Equations," John Wiley & Sons, Inc., New York, 1969.
2. Reddick, Harry W.: "Differential Equations," John Wiley & Sons, Inc., New York, 1949.

INDEX